The 7th (Queen's Own) Hussars

Harry Payne

SLARKADE 1864

The 7th (Queen's Own) Hussars

On Campaign During the
Canadian Rebellion, the Indian Mutiny,
the Sudan, Matabeleland, Mashonaland
and the Boer War

Volume 3: 1818-1914

C. R. B. Barrett

LEONAUR

The 7th (Queen's Own) Hussars: on Campaign During the Canadian Rebellion,
the Indian Mutiny, the Sudan, Matabeleland, Mashonaland and the Boer War
Volume 3: 1818-1914
by C. R. B. Barrett

Originally published in 1914 under the title
The 7th (Queen's Own) Hussars

Published by Leonaur Ltd

Text in this form and material original to this edition
copyright © 2008 Leonaur Ltd

ISBN: 978-1-84677-520-8 (hardcover)
ISBN: 978-1-84677-519-2 (softcover)

http://www.leonaur.com

Publisher's Notes

The opinions expressed in this book are those of the author
and are not necessarily those of the publisher.

Contents

CHAPTER 1

Home Service
1818-1838

For a time, after much hard campaigning, the 7th Hussars would rest from bivouac and battlefield. For no less than twenty years the regiment was employed on home service. Few events of the slightest importance to the regiment took place. Riots, it is true, more than once caused their service to be requisitioned by the civil power: and almost immediately on their return the death of Queen Charlotte at Kew, and her funeral procession thence to Frogmore and Windsor, caused a detachment, if not the whole Regiment, to be ordered to attend the ceremony.

It will be remembered that on its return from France the Regiment was quartered at Chertsey and in the neighbouring places.

Queen Charlotte died on 17 November. On 30 November and 1 December the body lay in state at Kew Palace. At 10 a.m. On 2 December the funeral procession started for Windsor, halting at Frogmore, where the Prince Regent as chief mourner joined the *cortège*. The *Manuscript Regimental Record* states that 'they assisted in the ceremonial of the Queen's funeral, which took place on the 2nd December.'

Cannon tells us that ' On the night of the 1st of December the regiment attended the funeral of Her Majesty Queen Charlotte.'

If the *London Gazette* is to be taken as an authority, Cannon's date is wrong. Also the escort on the 2nd was composed of a 'Field-officer's guard, with a standard, consisting of 120 men of the 19th Lancers.' At Longford, where a halt of one hour took place, the lancers were relieved by a similar guard of the 3rd Regiment of

Dragoon Guards. At Datchet bridge these fell into the rear and formed the rearguard, their post being taken over by a field officer's detachment of one hundred men, with a standard, from the Household Brigade. St. George's Chapel was reached at 7.30 p.m. The escort of the Prince Regent was from the Life Guards.

Troops, however, both of cavalry and infantry, lined the road from Frogmore to St. George's Chapel, and among the former it would appear that the Regiment was included, unless they acted as guard of honour during the preceding night outside Kew Palace.

We are told that every sixth man carried a *flambeau*. We are also told that the royal coffin was borne into the chapel from the hearse by ten Yeomen of the Guard. As a matter of fact, a specially constructed bier or car on six small wheels, the device of Sir William Congreve, was for the first time employed. It appears that the weight of royal coffins had in the past proved very excessive. One of the bearers after the funeral of the Princess Charlotte actually died in consequence of an injury he sustained in the performance of this duty. Moreover, too, the need of changing the bearers during the procession from the door to the choir had proved over and over again to be most inconvenient: hence Sir William Congreve's car.

Curiously enough, though the *London Gazette* gives the number of the Lancer Regiment as 19, the *Gentleman's Magazine* states that it was the 16th Lancers who escorted the body to Longford.

The *Annual Register* gives a lengthy account of the ceremony, from which we gather one or two pieces of information. While praising the general conduct of the ceremony and the excellent arrangements by which those present were able to separate without confusion and in a very brief time, that periodical remarks on the 'limited attendance and homage paid by the peerage on this occasion, as well as by other persons who occupy a distinguished rank in the state.' The 'choir was by no means filled by the numbers who were in attendance.' The cavalry, it adds, 'without any exception, behaved with the utmost propriety, and performed their civil duties of maintaining order with much good temper and courtesy.'

There were complaints however, of the behaviour of another branch of the service towards the crowd. Some criticism is bestowed on the selection of the troops present, and the *Annual Register* states that 'it was not apparently well suited to the occasion.'

One incident is really extraordinary. The procession arrived at George's Chapel 'nearly an hour sooner than it was expected; and some inconvenience resulted from this unusual excess of punctuality.' matter of fact, a large number of distinguished persons were shut out in consequence. Among them were the Chancellor of the Exchequer, Lord Bathurst, and the Duke of Northumberland (one of the pall-bearers). Some of the higher officers of the late Queen's household with the greatest difficulty succeeded in obtaining admittance. The ante-chapel, save for a few soldiers with *flambeaux*, was absolutely empty, or near so. The whole ceremony as a spectacle would therefore appear to have been mismanaged.

After the funeral the Regiment marched to Staines. Here they remained in quarters until 1 January 1819, when they marched for Manchester. From the *Morning Chronicle* of 23 June 1819 the following curious piece of information is derived:

> Whilst the 7th Hussars lately passed by the Marquess of Anglesea's seat near Lichfield, on their route to Manchester, they were entertained by the Marquess at his Mansion with good old English cheer. Whilst the soldiers were parading on the lawn in front of the house, immediately before their departure, a somewhat singular appearance presented itself in the persons of the Marquess, his brother (a Captain in the Navy), Lord Uxbridge (the Marquess's son), and the daughter of the Marquess. The first wanted a leg, which he lost at Waterloo, the Captain an arm, the Noble Lord Uxbridge was on crutches, being wounded in the knee, and the fair Lady was minus her right hand, which she lost while attending her husband at one of the battles in Spain.

Remaining at Manchester until 9 June, the Regiment was then ordered to Scotland. The events of this year are thus recorded in the MS. Book:

> At different periods the Headquarters were stationed at Hamilton, but were frequently moved from one station to another in consequence of the disturbances which took place in Glasgow and the neighbourhood, and finally one squadron was permanently stationed at Glasgow and one troop at Paisley until the Regiment left the country.

It will be remembered that it was in this year that the so-called *Field of Peterloo* affray took place at Manchester (16 August). On the evening of that day there were most serious riots at both Glasgow and Paisley. The *Gentleman's Magazine* records the event in the following terms: '*Sept.* 19th.—The accounts from the North state, that at Glasgow and Paisley, Reform meetings have been held; and disorders have taken place, under circumstances at once criminal and disgraceful. We fear they were not sudden ebullitions of popular excesses, provoked by real or imaginary injustice, but a determined and pre-concerted spirit of attack upon the public peace and private property.' The riot at Paisley is thus described: 'A meeting on Mickleriggs-Muir, near Paisley, having been advertised for September nth the sheriff of Renfrew and provost and magistrates of Paisley, issued a proclamation declaring against the avowed intention of bands of persons from different parts on going to and from such meeting, to parade the town of Paisley *with flags and devices of a political and inflammatory nature*; and warned all who should take part in such *illegal* proceedings, that they would be made responsible for their conduct. This notification was utterly disregarded by the reformers, who, after the meetings, came marching in great force, with music sounding and flags flying through the High Street of Paisley. The magistrates caused the colours to be seized, and in consequence a violent disturbance began: lamps and windows were broken, and special constables maltreated. The sheriff, the provost and magistrates who went among the mob to advise them to disperse, were assaulted with stones; and it was not till a late hour that order was restored by the reading of the riot act, and the apprehension of about twenty of the ringleaders. Similar outrages were renewed during several following days, in which several houses were gutted, many persons abused, and some robbed, by the plunderers, who took advantage of the occasion. After considerable forbearance, as well as all exertion on the part of the civil power, some cavalry (7th Hussars) were sent for from Glasgow, by whom the streets me repeatedly cleared By such means the spirit of outrage was at length subdued, and fortunately without the loss of a single life, though many persons were severely wounded; some by the mob, and some by the soldiers.'

It is curious, in connection with these disturbances, to read the following letter issued by the Home Department:

Whitehall
November 6th
My Lord, Having been informed that there are laying about throughout the kingdom, especially in the maritime parts of it, a great number of cannon which are private property, a considerable part of which were formerly used in merchant's ships, I beg leave to call your Lordship's attention to this subject; and to request that you will direct the magistrates of the county under your Lordship's charge, to make the necessary inquiries within their respective districts, and if any guns of this description should be found therein, that they will cause *immediate* steps to be taken with the consent of their owners, for rendering them useless, or for removing them to a place of security.

I have the honour to be, &c. &c.

Sidmouth

H. M. Lieutenant of——.

The Regiment remained in Scotland until 17 August 1820, when having marched to Port Patrick, it embarked there for Ireland, and landed at Dundalk on the 21st. Here the headquarters were established, detachments being sent to Belturbet, Monaghan, Enniskillen, &c.

It must be remembered that at this time in Ireland there were very few barracks capable of accommodating an entire regiment. The country was however, dotted with small barracks. Of these, some would hold a troop, more only half a troop, and occasionally an entire squadron could be housed. In this respect it was unlike England. There existed the practice of billeting on the inhabitants and in the inns where barrack accommodation was wanting, and it was not until the cavalry-barrack building era set in, i.e. from 1794 to 1808, that the billeting above mentioned was gradually dropped. In Ireland public-houses were many, but were most unsuitable in the country districts; and except in towns the houses of the people were of such a character as to be unavailable for the purpose.

Hence we find that in the case of a long stay in Ireland a regiment would perhaps occupy nearly every barrack in the country in detachments of various sizes. Of one cavalry regiment it is recorded that during its service in Ireland it absolutely occupied every barrack in the country save that at Longford. And the Irish cavalry barracks of those days: what places they were in which to house troops! insanitary, dilapidated structures—some indeed on the point of collapsing. One frequently finds mention in the official lists, of —— *barrack: since tumbled down*, or *unfit for occupation, to be repaired*, and that some years after the condemnation of the building.

In the large towns it was different, and cavalry were then housed in comparatively comfortable quarters. During this year George III died; but the Regiment, being at that time in Ireland, was not represented at the royal funeral—at least, the accounts of the procession do not record it.

The coronation of George IV took place on 19 July 1821. Being in Ireland, the Regiment was not engaged in the military portion of the ceremony, neither was it concerned in the funeral of Queen Caroline, on which occasion it will be remembered there was a considerable disturbance and several people were killed or wounded.

On 11 August George IV started on his visit to Ireland, arriving at Howth Pier about 4 p.m. the same day. No troops were stationed either on the pier or on the road when the King landed, nor does there appear to have been even a guard of honour mounted. After his arrival at the Viceregal Lodge the guns fired a royal salute and a lancer regiment and two companies of the Grenadiers entered the Park.

On 17 August the King made his public entry into Dublin. His escort was furnished by the 13th Light Dragoons. Next day the 7th (Queen's Own) Hussars, who had been marched to Dublin for the purpose, were reviewed with the rest of the garrison by the King. The review took place on the usual ground in the Phoenix Park. On the occasion the Marquess of Anglesea commanded the Regiment.

Reductions in the strength of the regiments of both infantry and cavalry now took place, and accordingly on 6 September the establishment of the 7th Hussars was reduced from eight to six hoops. The Regiment appeals to have remained in Dublin until 5 December 1822, when it marched to Newbridge and remained there until May 1823, when its quarters were shifted to Cahir, a troop being detached to Newross, and one to Filkard: evidently New Ross and Fethard. In June the Regiment reunited at Waterford and there embarked for Bristol. On landing it marched towards London, being quartered at Richmond and in the neighbourhood.

Here they remained until 15 July, when a review of the cavalry was held on Hounslow Heath. After the review the Regiment marched for Brighton, detaching a squadron to Chichester and a troop to Hastings. These detachments furnished parties along the coast who were engaged on the revenue duty.

In March 1824 the Regiment left Brighton, marching for London, and was stationed at Hampton Court and the neighbouring villages. Here they were engaged on brigade field days prior to the general cavalry review which took place on 14 July.

After the review the headquarters marched into Hounslow Barracks, but one squadron was left at Hampton Court and a troop was detached to Kensington. A party of thirteen men and twenty horses was also sent to the Royal Military College at Sandhurst.

It was the custom at one time for successive parties of cavalry to be rationed at Sandhurst in order to instruct the cadets in riding. The cavalry detachment for Sandhurst was invariably supplied from Hounslow until 1871. The Regiment remained at Hounslow until 4 July 1825, when it marched to York, where the headquarters were fixed. A squadron was, however, sent to Leeds, a troop to Newcastle, and another to Beverley. On 31 March 1826 the 7th were ordered to Scotland and marched from York, Leeds, Newcastle, and Beverley for Edinburgh, where they arrived on 13 April. One troop was detached to Perth to occupy the barracks there, and a party was sent on revenue duty to Cupar Angus and Forfar.

The stay of the Regiment in Scotland was less than a year, for on 1 March 1827 they returned to England, marching for Brighton and Chichester.

En route they were reviewed on Hounslow Heath by the Duke of Wellington on 12 April. The Duke was pleased to express his unqualified approbation of the appearance of the Regiment and the manner in which the different movements were performed.

One squadron arrived at Chichester Barracks on 18 April, and on the same day the headquarters marched into Brighton.

The 7th were now again sent to Ireland. They left Brighton and Chichester on 3 March 1828, marched for and embarked at Liverpool, and reached Dublin by 31 March.

Here they remained till 21 May 1829, when their quarters were shifted to Newbridge. After a stay at Newbridge for almost a year the headquarters were moved on 2 May 1830 to Dundalk, one troop being detached to Belturbet and one to Monaghan.

The Regiment does not appear to have been ordered to attend the funeral of George IV, nor to have been present at the coronation of William IV.

An important change took place in the uniform of the Regiment during this year.

On 2 *August,* 1830, a General Order was issued for the whole of the cavalry, with the exception of the Royal Horse Guards (the Blues), to be dressed in red. The 7th Hussars were consequently furnished with red pelisses in the following year. This point will be discussed later in Volume 4 on *Uniforms and Equipment.* But at the present moment it is sufficient to state that the blue pelisses were resumed in 1841.

In 1831 the 7th Hussars left Dundalk, Monaghan, and Belturbet, and returned to Newbridge, marching on 4 April. The Regiment was shortly afterwards ordered to leave Ireland, and leaving Newbridge on 23 June, marched to Dublin, where it embarked on 27 June for Liverpool, landing at that port on the following day.

From Liverpool the Regiment marched to Birmingham, where it arrived on 7 July. One squadron was detached and remained at Coventry and troop at Kidderminster. Here they remained until 8 March 1832, when orders were received to proceed to Norwich and Ipswich. Headquarters arrived at Norwich Barracks on 20 March. A squadron was stationed at Ipswich and a troop at Boston.

For a year the 7th Hussars remained in these quarters. On 28 March 1833, however, orders were received to proceed to Scotland. The Regiment accordingly marched north.

Headquarters proceeded to Hamilton Barracks, and a squadron was posted at Glasgow. On 15 July the headquarters removed to Glasgow. Hamilton was then occupied by one troop with the recruits and remount horses.

The *Manuscript Regimental Record* tells us that in the month of February 1834—

.... serious riots took place in Campsie, Dumbarton, Irvine and several other places in the vicinity of Glasgow amongst

OFFICERS' MESS

the calico printers and cotton spinners, in consequence of which the Regiment had pretty severe duty to perform. One troop was stationed at Dumbarton at 2nd Feb. to 26th Mar.

We are not told what the *pretty severe duty* was. From the *Gentleman's Magazine* however, we gather that for some time the state of affairs in Glasgow and the neighbourhood had been very serious. Attempts at assassination had taken place. There were also cases of arson, the latter luckily thwarted. Vitriol had been thrown through windows and more than one person was deprived of eyesight

As late as June of that year we read that in Glasgow at present, the Spinners' Union 'is in full vigour, and the employers are held in complete control.' It would not appear, however, that the Regiment was called upon to charge any riotous mob; but that their duties consisted in wearisome guarding of buildings and constant patrols in districts where able was threatened, or where a wholesome display of force was likely to act as a preventative of crime.

On 1 May the Regiment removed from Scotland, its destination being York. Headquarters arrived at that city on 15 May and three troops reached Newcastle on the 20th, under the command of Lieut.-Colonel Charles J. Hill.

On 20 April 1835 the Regiment left York and Newcastle. Headquarters arrived at Nottingham on 25 April, a squadron at Sheffield on 23 April, a troop at Derby on the 27th, and another at Boston on 2 May.

The Boston troop remained at that place until 31 October, when it was recalled to headquarters.

In April 1836 the Regiment marched under the command of Lieut.-Colonel C. J. Hill—headquarters and four troops being posted at Hounslow and a squadron at Hampton Court. Parties were detached to Kensington and Sandhurst from headquarters.

During the month of January 1837 an unpleasant adventure befell a captain in the Regiment. In the report of the affair the name is not given. One of the troops was stationed at Hampton Court, and the officer in question had occasion to drive thither on duty from the headquarters at Hounslow. He accordingly started in a tilbury, attended by a groom and driving a horse that was blind. The outward journey was accomplished without misadventure, but on returning at midnight the horse ran into a deep snow-drift, and started plunging violently, with the result that the driver, the groom, the horse and the tilbury were speedily overturned in a ditch at the side of the road. The weather being bitterly cold the officer had prudently wrapped himself up in sundry greatcoats and pea-jackets, and with a grazed right leg, an injured hip, and a damaged shoulder, was utterly unable to extricate himself from his uncomfortable position. The groom, who had escaped damage, however, managed to pull his master out and then set forth to obtain assistance. Having discovered a doctor, that worthy at first came to the conclusion that the sufferer was in a condition produced by having dined not wisely but too well. Further investigation, however, caused him to modify his first and hasty diagnosis. Luckily a piquet of the Regiment had been sent out by the commanding officer to traverse and watch the roads. The damaged captain was found by them with the attendant medico and carried back to the barracks, where his injuries were properly attended to.

The Regiment was now again about to proceed to Ireland.

On 28 May 1837 it marched in two divisions, under the command of Lieut.-Colonel J. J. Whyte. The first division consisted of headquarters and three troops, which embarked at Bristol on 6 June and landed at Cork on the following day. The second division embarked on the 12th and disembarked on the 14th. The Regiment was distributed thus: Headquarters and four troops at Ballincollig and a squadron at Cork. On 17 August, however, headquarters were removed to Cork

The Regiment was now about to proceed to Canada. In accordance with orders received the 7th (Queen's Own) Hussars marched for Dublin on 11 September 1837 in three divisions. They arrived at Portobello Barracks on 2 and 26 September and 3 October, where they remained during the winter. They were then formed into four service and two depôt troops.

The service troops, under the command of Lieut.-Colonel J. J. Whyte, marched from Dublin for Cork on 1 April 1838, the two depôt troops being left under the command of Captain Bushe.

It may be remarked that in January, Major Biggs and Veterinary-Surgeon Johnson, accompanied by three privates, had already proceeded to Canada to purchase horses for the Regiment.

We have elsewhere remarked on the small numbers of either

DUBLIN, 1837

officers, non-commissioned officers and men of Irish nationality in the Regiment. But it is not uninteresting to consider the probable cause: That is apparently due to the fact that the visits of the Regiment to Ireland had been few and comparatively brief. Between 1690 and 1838 we find that only five times had the Regiment been stationed in that island. The dates are 1713-14, 1810-13, 1820-23, 1828-31 and 1837-38. Why the Regiment, when not on active service, was thus almost entirely confirmed to England and Scotland it is hard to say, especially when we find other cavalry regiments treated in so different a manner. Cases have been known when long years were passed by regiments in Ireland. In one case a cavalry regiment passed over forty years in various billets in Ireland, and was scattered over the face of the land—a troop here, a squadron there, and a half troop elsewhere. It may be easily imagined it such a course did not conduce to anything approaching parade-ground proficiency except in very small numbers, and often indeed a lieut.-colonel-commanding never saw his regiment concentrated from one year's end to another. In these days regiments were placed on at was called the Irish Establishment, were paid at a different rate to those on the British Establishment, and differently governed also. The 7th Hussars, as will be seen, till the 19th century, passed only one year in Ireland, and hence it is, we conclude, that Irishmen in any numbers were not to be found in its ranks.

CHAPTER 2

Canada

1838

The service troops embarked for Canada as follows: on the *Arab* transport thirty-eight men and twenty-seven horses, 1 May. On the *Elizabeth* transport thirty-eight men and twenty-seven horses, and on the *Vanilla* transport thirty-six men and twenty-six horses. Headquarters and the dismounted men was the last division to leave England, and sailed in the *Prince Regent* transport on 12 May. Officers were permitted to take two chargers each. The transports arrived at Montreal on 8, 14, 20 and 24 June respectively. The men and horses were landed, and by 20 August the horses purchased in Canada having arrived, the Regiment was mounted to its full establishment. On the passage the following casualties occurred: One private, two horses and two officer's chargers died on board the *Arab,* and two horses on each of the other vessels.

1838.—In *The Naval and Military Gazette and East India and Colonial Chronicle* (now *The Broad Arrow*) for 14 July we read the following:

A correspondent writes: The arrival of a detachment of the 7th Hussars at Montreal occasioned considerable excitement in the city, and immense crowds were on the island wharf all day witnessing the landing of the horses—a novel importation from England to Canada. Considering the length of the voyage the horses looked well and excited very general admiration. The men all wear moustaches, which makes them look formidable.

On 27 June two troops were stationed at Chambly, but on the arrival there of the 1st Dragoon Guards from Three Rivers to occupy that station they were removed to Laprairie on 17 October. *The Manuscript Regimental Record* tells us as follows:

> On Nov. 4th the second revolt of the Canadians took place, and the rebels being in considerable force at the village of Napierville of which they had taken possession and imprisoned the loyal inhabitants, the Regiment with the other troops under Lieutenant-General Sir John Colborne marched upon that place on the 7th, and on the morning of the 9th entered the village of Napierville when it was found that the rebels had abandoned it about one hour previous to the arrival of the troops; setting their prisoners at liberty. After scouring the adjacent country for some days, and the rebels being totally dispersed, the Regiment returned to Montreal on Nov. 30th, leaving as before a squadron at Laprairie.

Sir John Colborne was a distinguished soldier. He had entered the army in 1794. Among his war services may be named the Helder (1799); Egypt (1801); Sicily where he was present at the battle of Maida. He was with Sir John Moore in Sweden and in the Coruña campaign as military secretary. Already a brevet-major, it was through Moore's dying request to Colonel Paul Anderson that Colborne received a brevet lieutenant-colonelcy.

He returned to the Peninsula in 1809, and was present at Busaco, Albuera. the storming of San Francisco (Ciudad Rodrigo). Salamanca. Vittoria, the Nivelle and the Nive as a brigadier. Colborne commanded his regiment at Orthes and Toulouse.

Promoted colonel on 4 June 1814, he received a gold cross and three bars. He was one of the first K. C. B.'s, and was afterwards *aide-de-camp* to the Prince Regent. At Waterloo, where he greatly distinguished himself, he commanded his regiment (the 52nd). His flank volley against and charge of the Old Guard is well known.

Colborne received the orders of Maria Theresa and St. George, and was promoted major-general in 1825; and at once made Lieutenant-Governor of Guernsey. In 1830 he was appointed Lieutenant-Governor of Upper Canada, an appointment he vacated on his promotion to the rank of lieutenant-general in 1838. At the moment he was preparing to leave Canada the rebellion broke out.

Colborne was then ordered (if he had not embarked) to assume the office of Governor-General and Commander-in-Chief. He speedily quelled the rebellion and acted with such prudence and promptitude under difficult circumstances that he was rewarded by a peerage, his title being Lord Seaton of Seaton in Devonshire. Lord High Commissioner of the Ionian Islands for 1843 to 1849, he was made G. C. M. G. In 1854 he was promoted general. From 1855 to 1860 Commander of the Forces in Ireland and Privy Councillor in that country. Lord Seaton retired in 1860, and was created a field-marshal. He died in 1863, aged eighty-five.

The story of the Canadian rebellion is as follows *(Gentleman's Magazine,* 1839):

> The Canadas have again been the scene of rebellious war and piratical invasion; the rebellion in the Lower Province, and the invasion in Upper Canada.
>
> The Rebellion began in the sub-district of Beauharnois, Chateauquay, and Acadie, occupying the western angle formed by the confluence of the Richelieu and the St. Lawrence. On the third of November an attempt was made to seize and burn the steamer *Victoria,* which had carried some artillery from Montreal. The vessel escaped, but the rebels possessed themselves of the town of Laprairie without opposition, and compelled the loyal inhabitants to ship themselves for Montreal.
>
> On the following day the rebels attacked the village of Beauharnois, and after a short but sharp conflict obtained possession of it. In the house of a gentleman named Brown, they captured Mr. Edward Ellice, a member of the Imperial Parliament, and nephew of Earl Grey, with his lady and her sister, and several others; whom they sent to Chateauquay, where they were secured in the house of the Roman Catholic clergyman.
>
> The rebels established their headquarters at Napierville, and their forces mustered, at one time 8000 men, generally well armed. They spent successive days at this town in the indulgence of the wildest excess. Meanwhile Sir John Colborne proclaimed martial law on the 4th Nov. and on that day a tribe of Caughnawagas Indians attacked and defeated a numerous body of the rebels, and made 75 of them prisoners. On the 8th Colonel Taylor, and a body of 200 British settlers, defeated five times their number of rebels and United

States' allies, marching to join the main body at Napierville. The engagement took place at Odellton, within sight of the United States' frontier. The rebels and their auxiliaries were commanded by Doctor Nelson, one of those excluded from mercy by Lord Durham's ordinance. On the 9th the rebels broke up from Napierville, and on the 10th and 11th Beauharnois and Laprairie were retaken. Sir John Colborne concentrated his troops on the 12th at Napierville and Chateauquay, and executed according to the Montreal journals, a severe vengeance upon the rebels whom he found there, burning the houses of the disaffected through the whole district of Acadie. This melancholy but unavoidable visitation upon the guilty terminated the rebellion.

An invasion at Upper Canada, by which the attention of the British Commander might be distracted, and the military force of the colony divided and weakened, was however, part of the concentrated (concerted?) plan of the traitors and their republican confederates. Accordingly, on the night of the 11th, at the moment when at the distance of about 180 miles eastward, Sir John Colborne was putting the last hand to the suppression of the rebellion in Beauharnois and Acadie, 800 republican pirates embarked in two schooners at Ogdenburgh, fully armed, and provided with six or eight pieces of artillery, to attack the town of Prescott, on the opposite side of the river. They failed in the attempt to disembark at Prescott, but by the aid of two United States' steamers, effected a landing a mile or two below the town, where they established themselves in a windmill and some stone buildings, and repelled the first attempt to dislodge them, killing and wounding forty-five of their assailants, among whom were five officers: but on the 15th, Colonel Dundas brought a reinforcement of regular troops, with three pieces of artillery, against the invaders. From the water the rebels were fired upon by Captain Sandom, who had two gunboats; and, after enduring the attack for about an hour, they hung out a flag of truce and surrendered at discretion.

At Caughnawaga, by order of Lieut.-General Sir John Colborne, Captain Campbell, of the 7th (Queen's Own) Hussars, was attached to the Indians. Probably he acted in the double capacity of commander and also to prevent scalping. On 3 January 1839 a report reached Montreal that a body of rebels had assembled at Ferrebonne. One troop of the Regiment under the command of Captain Campbell was therefore despatched thither. Nothing, however, occurred, and the troop returned to Montreal on 5 January. During the year 1839 the Regiment for six weeks went into camp on Laprairie Common for the purpose of field exercises. On conclusion of the exercises they returned to their former quarters, where they remained during the rest of their stay in Canada, i.e. 1840-41 and part of 1842.

The Captain Campbell alluded to as attached to the Indians at Caughnawaga had a distinguished and diversified career. Details of his services will be found in Appendix 2.Volume 4. By the kindness of Edmund A. Campbell, Esq., his grandson, who has forwarded to me several commissions, letters, a notebook, and a brief biography, I am able to insert one or two very interesting regimental details.

Dated 'Headquarters Casseres, 2nd March 1814' is cavalry order signed by Major-General Lord Edward Somerset, who it will be remembered at that time commanded the cavalry brigade of which the 7th (Queen's Own) Hussars formed a part. It runs as follows, and refers to the battle of Orthes:

> Major-General, Lord Edward Somerset has much pleasure in making known the cavalry orders of the 28th Feb. and the 1st and 2nd of March, to the corps, composing his brigade and joins with Lieutenant General Sir Stapleton Cotton in expressing his perfect approbation of the conduct of the brigade in the battle of the 27th ultimo. He has particularly to thank Col. Kerrison and the 7th Hussars, as also Major Gardner and his troop of Royal Horse Artillery, who had an opportunity of being more closely engaged with the enemy on that day.

There is an interesting letter to Captain Campbell from E. Cotton, of Waterloo Museum fame, who it will be remembered had served in the 7th Hussars. Cotton also enclosed a copy of Lord Uxbridge's letter to the officers of the Regiment regarding Genappe, which we have already given *in extenso*. Here we shall quote a portion of Cotton's letter:

Mont St Jean
Waterloo
Belgium
26th May 1837

Sir, I hope you will excuse the liberty I take, in writing to you, as it has struck me frequently that the general cavalry order respecting the conduct of the 7th Hussars at the Battle of Orthes may not be in the records of Regiment.

Cotton was quite correct: the order does not appear, and we have gladly inserted it above.

I take the liberty of sending you a copy of it; also a copy of Lord Anglesey's letter to the Regiment on the affair near Genappe, the latter I am often obliged to show to visitors here, to defend the character of my old Regt., of whose honours I am as jealous as if still in its ranks, and shall defend its conduct against all assailants.

Sir, I have this day purchased one of our Regiment's rifle carbines, that was left on the field, at the battle, it appears to have been lying by and never used, or cleaned, from the day of action. It is marked 7 L. D., A. 11. Mr Blackier no doubt can tell to whom it belonged, probably the officers would wish to have it, if so, I will forward it to the Regt. On receipt of a line to that effect.

Thomas Blackier was the regimental quartermaster in 1810. He had been present at Waterloo and died or retired in 1859. He could of course easily have traced A.11. No application appears, however, to have been made to Cotton for the carbine. The letter continues:

I have ascertained the name of the farm that Sir E. Kerrison Bart., and Sir H. Vivian Bart, quartered at on the night of the victory of Waterloo. It is called Hallincourt, about 4 miles in advance of the British position and one mile to the right of the Charleroi Road. . . .

The anniversary of Waterloo is drawing near, but I am sorry to say there is no review or any other procession here on that day, in fact since the Belgian Revolution the day appears forgot. I was near getting into a serious scrape last 18th of June for hanging black as mourning on the monuments, the *gens d'armes* said such things may cause a disturbance, but the Mayor of Brainlallude in whose commune it is, told me he was glad I had shown such respect to the fallen heroes, and I may do it every year, which I shall not fail to do if I live here for 50 years to come.

Sir, if there is any officer in the Regt. acquainted with the Earl of Aberdeen, it would be well to inform his Lordship of the dilapidated state of the monument of the late Lieut.-Col. Sir Alexander Gordon. I am quite certain Lord Aberdeen is deceived by some person, as he has given orders for the repairs of the monument, but it has never been done, and if left much longer will fall completely down.

A rather amusing song—a parody on the old and then popular *Fine Old English Gentleman*—is also among the Campbell papers. It is dedicated to 'Capt.——, 7th Hussars by A Subaltern.' It seems that one of the captains in the Regiment lacked smartness in the matter of his uniform, and was inclined to practise sartorial economy on an extended scale. But let the song tell its own story:

I'll sing you a good old song which I have heard of late,
Of a fine old regimental coat, whose birth is out of date,
But still there is a strange report, which may be false or true,
That nine good years have recognised the everlasting blue,
Of this fine old regimental coat, all of the olden time.

This coat so old was patched around with various bits of cloth,
And here and there, by little, the original shone forth.
Twas here a piece from Birmingham proclaimed the tailor's art,
And Norwich too and Manchester could each display their part
Of this fine old regimental coat, all of the olden time.

The custom was when lady fair or general might appear,
To overhaul the Regiment, which happens once a year.
This fine old coat pro tempore was thrust into the store[1]
Great pity 'tis but doctors knew the renovating power,
Of this fine old regimental coat, all of the olden time.

But all, alas! must bend to fate, on one unlucky day,
The tails from the old body with one shrill crack pass'd away,

---

1. Should not this line have run 'was drawn forth from the store'? *Power* and *store* do not rhyme.

The captain's tears right mournfully bedewed the tattered shade,
And where's the coat, now² has been seen so often on parade,
As this fine old regimental coat, all of the olden time.
But tho' pelisses soon wear out and jackets pass away,
Economy in dress it seems has prov'd our captain's stay,
For though his back may not be clothed as it was wont of yore,
Yet soon I guess he will produce a rival from his store,
Of this fine old regimental coat, all of the olden time.
Tally hi hi ho—Yoicks!

In a little red leather covered notebook, which at one end contains information as to the B Troop of the 7th Hussars and at the other similar entries as to the 2nd Dragoon Guards, we find the following facts: in 1832 the average age of men of the troop was twenty-seven years and seven months; of the Regiment thirty years and one month. B Troop contained forty-six Englishmen, three Scots, and five Irishmen;³ the Regiment two hundred and fifty-six, thirty-six, and forty-three. Average height: 5 feet 9¼ inches in the troop; 5 feet 9½ inches in the Regiment. Horse: 15 hands 1¾ inches in the troop, 15 hands 2 inches in the Regiment. Age of horses: six years in the troop; seven years eight and a half days in the Regiment. Religions: Protestants fifty-three, Roman Catholics one, in the troop; three hundred and fifteen and twenty in the Regiment. Women and children: nine women and thirteen children in the troop; sixty-three and ninety-two in the Regiment. Then follows a price fist of necessaries; and directions how to pack the valise, the length of the latter being given as thirty inches.

METHOD OF PACKING THE VALISE

Near side: 1 shirt, 1 towel, 1 pair of drawers, 1 flannel jacket, 1 pair of socks rolled in white trouser.
Off side: 1 shirt, 1 pair of socks, 2 brushes, 1 pair of gloves, 1 towel, trap case, rolled in blue overalls. a pair of shoes (heels inwards) in the flaps, with one brush in each and blacking box in either, stable jacket under the flap with the plume and

2. Query for *now* read *that.*
3. We have elsewhere drawn attention to the preponderance of men of English birth over Scots and Irish in the ranks of the 7th (Queen's Own) Hussars.

27

button stick inside. Collar chain in the off holster with the curry comb and brush, trap bag, corn bag and stable cap on the top of the holsters. shoe cases buckled round the hind fork on each side so as to be flat on the blanket, just below the saddle bars.

The cloak was to extend for two inches below the holster pipe. Next is a complete list of returns required from troops when at out quarters, stating when they are to be made up to, when to arrive, and to whom addressed. The average weight of the men and accoutrements is given as 18st. 3lbs.

A complete muster-roll of the B Troop for 1832 follows, giving age, arms number, horse number, when attested, and height. For the horses we have date of joining and age; but this is not complete.

A list of the front rank is given, but that of the rear rank is missing. In pencil on the last pages are remarks on the characters of the men of the troop. Most of them are good, some indeed 'excellent in everything.' One man is 'occasionally given to drink,' one 'not very smart,' one 'very clean but a little given to drink,' two are 'very drunken' and two 'very bad soldiers, very drunken.'

It is not generally known that cricket grounds were constructed for the use of troops at the respective barrack stations throughout the United Kingdom as early as 1841. The grounds were placed under the charge of the barrack masters, but under the authority of the officer commanding the station. To wantonly damage the ground was to be considered as a *grave offence*.

The troops were ordered to pay for any repairs. This is a curious little piece of information, and from the fact that it shows that as far back as 1841 the authorities endeavoured to foster games, even though to a limited extent only, is certainly worth record. The writer discovered it in the War Office library, and it is dated 2 March 1841.

The Regiment received orders to return to England during September 1842, and sailed in the months of October and November. During the absence of the service troops in Canada, the depot troops at home had been stationed at Dublin, Leeds, York, Dorchester, Weymouth, and finally at Canterbury.

CHAPTER 3

Home Service Again
1842

1842.—The Service Troops received orders to return to England, and embarked in five divisions as follows, *viz*.: 106 men, 12 officers' horses and 63 troopers left Montreal on 19th October and embarked at Quebec in the *Nautilus* transport; 53 men, 4 officers' horses and 34 troopers in the *Sovereign;* 53 men, 8 officers' horses and 29 troopers; 30 men, 2 officers' horses, and 28 troopers left Montreal on 1st November, and embarked at Quebec on the *Tyne*; 40 men, 2 officers' horses and 34 troopers left Montreal on 3rd November and embarked at Quebec in the *Tanjour;* 54 men, 7 officers' horses, 33 troopers left Montreal on 4th November and embarked at Quebec on the *Tory*.

The *Sovereign* and *Nautilus* arrived at Portsmouth on the 20th *November* and the *Tory* on the 17th December. The *Tyne* arrived at Ramsgate on the 19th December and the *Tanjour* on the 23rd; the whole joined the depôt troops stationed at Canterbury. The casualties which occurred were 1 horse left at Quebec, there being no room in the ship, in the *Tanjour*. Three horses died on the passage and 2 were destroyed on suspicion of glanders in the *Tyne*. 2 horses died on the passage in the *Sovereign*. 1 man and 1 horse died on the passage and 1 was destroyed after disembarkation at Portsmouth in consequence of injuries received whilst at sea in the *Nautilus*.

April, 1843.—The Regiment marched from Canterbury under the command of Lieutenant Colonel Whyte for Brighton and Chichester as follows, viz.: 1 squadron on the 19th and Headquarters with 1 squadron on the 20th for Brighton, and 1 squadron for Chichester on the 21st April. The squadrons stationed at Chichester marched for Dorchester in two divisions, viz.: 1 troop on the 30th June, and 1 troop on the 3rd July. A detachment consisting of 1 sergeant and 14 rank and file under command of Lieutenant Preston marched from Brighton for Chichester on the 7th August. One Troop under command of Lieutenant Wyndham marched from Dorchester for Trowbridge on the 10th August.

1844.—The detachment stationed at Chichester rejoined Headquarters at Brighton on the 19th March. The Regiment marched from Brighton under the command of Lieutenant Colonel Whyte in the following order viz.: 1 troop on the 15th April for Dudley, Headquarters and 2 troops on the 17th for Birmingham, and 1 troop on the 19th for Coventry. 1 troop joined Headquarters at Birmingham from Trowbridge on the 23rd April.

1 May, 1844.—1 troop under command of Captain Shirley marched from Birmingham to Stone in Staffordshire.

31 May, 1844.—1 troop joined Headquarters at Birmingham under command of Captain Grasett from Dorchester.

3 September, 1844.—1 troop joined Headquarters from Stone.

23 April, 1845.—1 troop joined Headquarters from Dudley.

24 April, 1845.—The Regiment marched from Birmingham under the command of Lieutenant Colonel Whyte for Ipswich and Norwich, detaching 1 squadron to Coventry in the following order, viz.: 1 squadron marched on the 23rd for Norwich, and 1 squadron with the Headquarters on the 24th.

The Squadron detached at Coventry marched on the 20th and 21st June, *viz.*: 1 troop for Norwich and 1 troop for Ipswich, and arrived at their respective stations on 30th June and 1st July.

April, 1846.—The Regiment marched from Ipswich and Norwich in the following order, viz.: 2 troops from Norwich for Hounslow on the 17th Instant under the command of Major Campbell.

1 Troop on the 18th under the command of Captain Shirley. Headquarters and three troops from Ipswich on the 21st under the command of Lieutenant Colonel Whyte, and arrived at Hounslow on the 25th of the same month. 1 troop arrived at Hampton Court on the 25th instant, 1 troop on the 27th, and parties detached at Kensington and Sandhurst, A painful and unpleasant affair took place in the month of July 1846, which in consequence of the public attention it occasioned needs to be recorded.

BRIGHTON, 1843

On 15 June a private of the Regiment at Hounslow who had been guilty of insubordination was sentenced to receive one hundred and fifty lashes. The punishment, it was proved, was not severely inflicted, if the subsequent evidence given by the doctor is to be credited, and there is no reason to impugn his testimony. The name of this man was Frederick White. It appears that for the purpose of punishment he was tied to a ladder, the ordinary triangle not being, it would seem, available; and as far as one can see this was the only departure from the ordinary method of inflicting corporal punishment.

After the flogging, White was as usual taken to the hospital. Here it was found at his back was not badly lacerated, the 'real skin not being cut through.' He was duly treated and all went well with him till the morning of 6 July, on which day he was to return to duty and be discharged from the hospital, his back being completely healed. White now complained of pain in the region of the heart, through his back and shoulder-blade. Dr. Warren, the surgeon of the Regiment, who had of course been present at the punishment, did all that he could to relieve the man. Paralysis of the lower extremities, however, was discovered, and the unfortu-

FREDERICK JOHN WHITE
PRIVATE IN THE 7ᵀᴴ QUEENS OWN HUSSARS
WHO DIED 11ᵀᴴ JULY 1846
AGED 27 YEARS

THIS STONE IS ERECTED BY HIS COMRADES
AS A TESTIMONY
OF THEIR SYMPATHY FOR HIS FATE
AND THEIR RESPECT FOR HIS MEMORY

RE-ERECTED
BY THE OFFICERS 7ᵀᴴ HUSSARS
MAY 1886

nate soldier died at 8.15 p.m. on 11 July. Another army surgeon was then sent by Sir James McGrigor, Bart., the Director: the Army Medical Department, to make a report on the affair.

A post-mortem disclosed that inflammation of the pleura and of the lining membrane of the heart existed, and that the man's death was 'in no way connected with the corporal punishment inflicted on 15 June.' An inquest was held publicly, during which it became apparent that the coroner was somewhat a partisan in his attitude towards the military authorities, medical and otherwise.

It was given in evidence that the man had, on 4 July, when seemingly in good health, been employed in cleaning out the mortuary and also other patients in his ward in the hospital on the ordinary duties performed by those in hospital.

Capital was endeavoured to be made out of the fact that the mortuary was a small room with a stone floor, and damp. It was sworn that during his punishment White made 'little or no motion with his body and kept his breast towards the ladder without the least struggle or twisting himself.' Dr. Warren stated that never 'did I witness so little muscular effort in all the punishments I have witnessed, as I did in this case.' There was ' no spasmodic action of the muscles of the back on either side, and being close by, if there had been I must have seen it.'

It was stated that on receipt of a letter from his brother in America on 7 July, White was 'observed to be despondent,' but how this affected the matter cannot be exactly discovered, as the serious symptoms appeared on 6 July. Public attention was of course drawn to the matter, and it came before the House of Lords on 14 August in connection with a petition for the immediate abolition of flogging in the Army. There was as usual much discussion, but no result.

The *Medical Times* appears to have attacked the coroner and a libel action followed, but this concerns us not. Attempts to abolish flogging in the Army were unsuccessfully made in 1876 and 1877. In 1879 flogging was reduced by the Army Discipline Act and rendered commutable to imprisonment. The total abolition of this form of punishment did not take place until 1881, when other penalties were substituted therefore. It is curious to note that there seems to be no record of the details of the offence for which this man was flogged.

According to the rules of the service at that date, the amount of punishment awarded was not stated to be excessive. A headstone, erected by the Regiment, on the grave of the unfortunate man still exists. He would appear to have been, as a soldier, popular in the Regiment.

21 September, 1846.—The first division of the Regiment, consisting of 3 troops under the command of Major Shirley, marched from Hounslow *en route* for Ireland *via* Liverpool, embarked at Liverpool on the 6th April and arrived at Dublin on the following morning, and proceeded for Athlone where they arrived on the 13th October. The second division of two troops under the command of Captain C. Hagart marched from Hounslow on the 24th and marched same route as first division. The third division consisting of Headquarters and a troop under the command of Cornet Bushe, Lieutenant Colonel Whyte taking command it Liverpool, where they embarked on the 12th October for Dublin and arrived the following morning, remained in the Royal Barracks for one night and marched the following day *en route* for Athlone and arrived on the 20th October, detaching troops as follows, *viz.*: 1 troop to Gort Captain C. Hagart), 1 troop to Galway (Captain G. T. Bushe), 1 troop to Ballinrobe (Captain C. L. Peel), 1 troop to Loughrea (Captain A. Helyar).

3 November, 1846.—1 detachment consisting of 1 sergeant and 10 rank and file marched to Birr Barracks.

20 November, 1846.—Cornet Bushe marched to take command of the detachment at Birr.

20 May, 1847.—Captain J. M. Hagart's Troop marched to Ballinrobe there to be stationed.

22 May, 1847.—Captain C. L. Peel's Troop marched into Headquarters in Ballinrobe

Lieutenant Colonel Whyte retired on half-pay the 16th April 1847 and Major A- Shirley promoted Lieutenant-Colonel the Regiment the same date.

11 September, 1847.—Captain Peel's Troop and detachment stationed at Birr marched this day to Dublin, arrived in Portobello Barracks on the 23rd of the same month.

23 September, 1847.—Captain Bowles' Troop marched from Gort for Dublin.

Captain J. M. Hagart's Troop marched from Ballinasloe for Dublin. Both troops marched from Athlone Barracks under the command of Captain J. M. Hagart and arrived in Portobello Barracks on the 1st October.

27 September, 1847.—Captain C. Hagart's Troop marched this day from Galway and arrived in Portobello Barracks, Dublin, on 6th October.

30 September, 1847.—Captain Sir W. Russell's Troop marched from Loughrea this day and arrived in Portobello Barracks, Dublin, on 7th October.

2 October, 1847.—Headquarters and 1 troop marched this day from Athlone under the command of Lieutenant Colonel Arthur Shirley and arrived in Portobello Barracks, Dublin, on the 8th of the same month.

23 June, 1848.—Captain Bowles' Troop marched from Dublin to Athy, in consequence of the horses being affected with glanders.

24 July, 1848.—Captain Bowles' Troop marched from Athy to Carlow.

Inspected by Major-General H.R.H Prince George of Cambridge, K. H., 7th October, 1848.

9 October, 1848.—Three troops of the Regiment marched from Portobello Barracks, Dublin, this day for the following stations, viz.: Captain J. M. Hagart's Troop to Waterford, Captain the Marquis of Worcester's Troop to Kilkenny, Captain E. H. Cooper's Troop to Carrick-on-Suir, detaching a party of 1 sergeant and 8 rank and file to Piltown.

12 October, 1848.—The Headquarters and two troops of the Regiment marched from Portobello Barracks to Newbridge this day under the command of Lieutenant Colonel Shirley.

13 October, 1848.—Captain Bowles' Troop marched from Carlow this day and arrived at Newbridge Barracks on Saturday, 14th October.

7 May, 1849.—The detachment which was stationed at Piltown joined Headquarters this day.

Inspected by Major-General H.R.H Prince George of Cambridge, 11th May, 1849.

13 July, 1849.—Captain Bowles' Troop marched from Newbridge for Waterford this day, and arrived at Waterford on the 15th instant.

16 July, 1849.—Captain Viscount St. Lawrence's Troop marched from Newbridge for Carrick-on-Suir this day and arrived on the 20th instant under the command of Lieutenant Sartorious.

Captain Cooper's Troop joined Headquarters from Carrick on 25th July.

17 July, 1849.—1 Sergeant, 1 corporal and 19 privates with 20 troop horses of Captain Sir William Russell's Troop marched from Newbridge for Carlow this day under the command of Lieutenant Cooke and arrived at Carlow on the 18th.

24 July, 1849.—Captain Hagart's Troop joined Headquarters from Waterford, 1 squadron under the command of Major Hagart marched to Dublin this day as an escort for Her Majesty during her visit to Dublin and rejoined Headquarters on the nth of the next month. During the stay of the Queen, the Regiment, with other cavalry regiments, was inspected by His Royal Highness the Prince Consort.

23 August, 1849.—The detachment which marched to Carlow on the 17th July rejoined Headquarters at Newbridge this day.

7 December, 1849.—Captain the Marquis of Worcester (F) Troop joined Headquarters this day from Kilkenny.

13 March, 1850.—Captain Hagart's Troop marched from Newbridge for Bandon this day and arrived at Bandon Barracks on the 23rd March.

1 April, 1850.—1 squadron of the Regiment (A & E Troops) under the command of Captain Sir William Russell marched from Newbridge for Ballincollig this day and arrived at Ballincollig Barracks on the 10th April, 1850.

2 April, 1850.—The Headquarters and 1 troop (F Troop) marched from Newbridge this day for Ballincollig under the command of Major Hagart and arrived at Ballincollig Barracks on the 11th April, 1850.

6 April, 1850.—Captain Bowles' Troop marched from Waterford this day to Cork and arrived in Cork Barracks on the 11th April, 1850.

9 April, 1850.—Captain Viscount St. Lawrence's Troop marched from Carrick-on-Suir this day under the command of of Lieutenant Sartorius for Cork and arrived in Cork Barracks on 12th April, 1850.

Inspected by Major-general Turner, 10th May 1850.

Inspected by Major-general Turner, 18th October 1850.

On this occasion the Regiment performed the following movements:

March past by squadrons; Rank past by sections of threes with carried arms; Trot past by squadrons and by troops. Canter past left in front.

On the Move: change position to the rear, by the echelon of troops. From the right of threes to the front file, at a canter, and perform the sword exercise.

The line will advance, and on the move retire by threes from the right of squadrons; Covered by skirmishers from the left flanks and heads of squadrons left; and half front, skirmishers in.

2nd and 3rd squadrons will advance in line, and attack to the front; supported in 2nd line by 1st squadron; after attack, halt, and 2nd and 3rd squadrons threes inwards, and form close column in rear of the 1st squadron.

Column will retire covered by skirmishers from right flanks of squadrons.

Column left incline, and halt, front, forward, and on the move deploy on centre squadrons by the incline of squadrons, and skirmishers in. Continue to advance, and on the move by echelon of troops change position half right back. Line will advance at a trot, and halt, change front half left on 2nd squadron.

Advance and on the move open column of troops to the left; change direction left, and line to the rear on leading troop.

Advance and attack in line; on the halt being sounded, the left troop will pursue along the whole front and reform in line round the rear.

Retire in echelon of squadrons from the left, covered by skirmishers of right division of right troop of line, Left division of do. to support.

Halt, and change direction to the left.

Halt, front, forward and line to the right on 2nd squadron. Skirmishers in, Advance in line, and on the move form double open column in rear of the centre.

Column will retire and form line to the front. Form open column of squadrons in rear of the right, or centre (as the case may be).

Attack to the front by successive squadrons, and retire to original ground; forming troops to the front on the move.

Column advance; on the move squadrons half left, and deploy on centre squadron.

Advance and on the move, left shoulders forward. Advance square across the ground; and on the move, at a trot, retire in open column from the right, and form close column to the rear, on the rear troop. Deploy on 1st squadron, and advance in line in parade order.

22 October, 1850.—Captain J. M. Hagart's Troop marched from Bandon to Cork Barracks, Captain Sir William Russell's Troop marched from Ballincollig to Bandon Barracks under the command of Lieutenant Cooke, and Captain Babington's Troop marched from Cork to Ballincollig Barracks; the whole took place this day.

21 January, 1851.—Captain Bowles' Troop marched from Cork to the Headquarters of the Regiment at Ballincollig this day under the command of Lieutenant Sartorius.

11 April, 1851.—Captain Sir William Russell's Troop marched from Bandon to Cork this day under the command of Lieutenant Cooke.

9 June, 1851.—Captain Hagart's Troop marched from Cork this day under the command of that Officer and Captain Cooper's and Captain The Marquis of Worcester's marched from Ballincollig this day under the command of Captain Cooper. The whole of these troops arrived at Island Bridge Barracks, Dublin, on the 18th Inst.

10 June, 1851.—The Headquarters and Captain Babington's Troop marched from Ballincollig this day under the command of Major Hagart and arrived at Island Bridge Barracks, Dublin, on the 20th June.

20 June, 1851.—The dismounted men marched from Ballincollig to Cork this day, thence by railway to Dublin the same day under the command of Paymaster Cubitt.

14 July, 1851.—Captain Sir William Russell's Troop marched from Cork this day under the command of Lieutenant Cooke, and arrived at Arbour Hill, Dublin, on the 23rd July.

15 July, 1851.—Captain Lord Garvagh's Troop marched from Ballincollig this day under the command of Lieutenant Bushe and arrived at Island Bridge Barracks, Dublin, on the 24th July.

LIEUTENANT JOSEPH TRENERRY & QUARTERMASTER JOHN EVANS PARRY, 1851

Inspected by Major-General H.R.H the Duke of Cambridge. 31st October 1851.

Lieutenant-Colonel Shirley retired on half-pay the 31st October, 1851, and Major C Hagart promoted Lieutenant-Colonel of the Regiment the same date.

From a *Field State* at Island Bridge Barracks, Dublin, dated 31 October 1851, we gather the following particulars. On parade, mounted, 1 field officer; 5 captains; 8 subalterns; 19 sergeants; 3 farriers; 197 rank & file; 20 officers' horses; 219 troop horses. The band consisted of 7 trumpeters, and 11 rank & file, with 18 horses. The total effective strength of the Regiment was 3, 6, 12, 6, 24; 7, 6, 320 respectively with 47 officers' horses and 266 troop horses. Wanting to complete:—1 sergeant; 2 rank & file and 5 troop horses. The establishment therefore was at the effective strength with the addition of the men and horses *wanting to complete.*

Under the heading 'Regimental Employ' we find 1 paymaster sergeant major; 1 sergeant saddler; 2 orderly room clerks (sergeants); 1 private; 1 mess waiter (corporal), 1 (private); 1 armourer (private); 1 acting schoolmaster (private); and 'At Work' 3 farriers and 2 privates. This is signed by Lieut.-Col. Shirley and is apparently the state of the Regiment on the date of his retirement, when the Regiment was inspected.

24 January, 1852.—A Squadron of the Regiment (C & F Troops) under the command of Captain Babington marched from Island Bridge Barracks, Dublin, this day and arrived at Dundalk on the 26th January, 1852.

28 *April,* 1852.—1 Squadron of the Regiment (B & D Troops) marched under the command of Captain Sir W. Cooke from Island Bridge Barracks and arrived at Dundalk on the 29th Instant, there to await embarkation for Glasgow.

29 April, 1852.—The young horses under the command of Lieutenant & Riding Master Brown marched from Arbour Hill Barracks, Dublin, and arrived at Dundalk on 1st May, 1852, there to await embarkation to Glasgow.

3 May, 1852.—The first division consisting of B, C, D & F Troops embarked under the command of Captain Babington at Dundalk for Glasgow, arrived and disembarked there on the 4th Inst: when B & D Troops under the command of Captain Sir W. Cooke proceeded

the same day to Hamilton there to be stationed; the remainder proceeded to Edinburgh where they arrived on the 6th Instant.

3 May, 1852.—Headquarters and A & E Troops marched under the command of Major Hagart from Island Bridge Barracks, Dublin, and arrived at Dundalk on the 4th Instant, embarked there with the young horses on the 6th Instant for Glasgow, arrived and disembarked there on the 7th Instant, and proceeded *en route* for Edinburgh (E Troop branching off at Glasgow and arrived there on the 10th Instant).

Inspected by Major-General Napier, 11th June, 1852.

Inspected by Major-General H .R. H. the Duke of Cambridge, 2nd October, 1852.

3 July, 1852.—Captain Sir W. Russell's Troop marched for Cupar, Fife, this day, in aid of the civil power, and returned to Edinburgh on the 6th instant.

Without further information we should be inclined to assume that this was an ordinary case of riot; and that the aid of the 7th Hussars had been requested to put an end to some trade dispute or political disturbance. It was, however, of quite a different kind, and the facts are these. In April 1852 an unfortunate old woman who was supposed to possess a little money was brutally murdered by two Irish labourers, brothers, by name Michael and Peter Scanlan. Another Irishman by name M'Manus was concerned in the crime but turned approver to save his life. The old woman was killed with blows from a three-legged stool, her head being terribly smashed. The brothers Scanlan were tried on 14 June at the High Court of Justiciary in Edinburgh. Being found guilty they were sentenced to death and remitted to Cupar for execution. The murder had taken place near Hilton of Fortha, where the condemned men, who worked in some lime works, lodged. The murder and condemnation caused great excitement in these remote districts. No execution had taken place in the neighbourhood for more than twenty years. A very large crowd assembled in the streets, the people having flocked in from considerable distances, and their attitude was the reverse of orderly. The magistrates therefore deemed it necessary to guard the scaffold with military and a large body of special constables. For this purpose, then, the troop of the 7th (Queen's Own) Hussars was summoned from Edinburgh. There cannot have been many more executions in Great Britain in which the ground surrounding the scaffold was guarded by cavalry. Hence the facts are worth recording.

1 June, 1853.—1 squadron of the Regiment A & E Troops under the command of Captain Sir William Russell marched from Piershill Barracks, Edinburgh, this day, and arrived at Hamilton on the 2nd June, 1853.

2 June, 1853.—One squadron B & D Troops under the command of Captain Sir W. Cooke marched from Hamilton this day, and arrived at Piershill Barracks, Edinburgh, on the 3rd June, 1853.

Inspected by Major-General H.R.H The Duke of Cambridge, 6th September, 1853.

Another programme of nineteen field movements written on a card is attached to the *State* and *Field Movements* already quoted. It is undated, but apparently belongs to this period.

No. 1.—Retire in column of troops from the left, covered by skirmishers of the right division of the right troop supported by the left division.

No. 2. —Line to the left about on the leading troop, skirmishers in and advance.

No. 3.—By echelon of troops change position left back.

No. 4.—Advance in echelon of squadrons from the left, and on the move the echelon will change direction to the right.

No. 5.—Advance and form line on the third squadron.

No. 6.—Advance and retire in open column from both flanks by the wheel about of threes and form double column in the centre, halt, front.

No. 7.—Line to the front.

No. 8.—Advance and attack in line; on the halt being sounded the right troop of the line will pursue across the whole front and reform in line.

No. 9.—Retire by alternate troops, covered by skirmishers of the left division.

No. 10.—Line on the right troop.

No. 11.—Advance in open column from the right and form line to the front on the 2nd squadron, and skirmishers in.

No. 12.—Advance and retire in columns of troops from the right.

No. 13.—Form close column to the rear on the rear troop.

No. 14.—Attack by squadrons in succession and retire to the original ground.

No. 15.—Column will advance and on the move deploy on the first squadron.

No. 16.—2nd squadron will dismount with carbines and skirmish. 1st squadron will cover them.

No. 17.—Skirmishers in and by echelon of troops change position half right back.

No. 18.—Advance and change front half left.

No. 19.—Advance in parade order.

31 May, 1854.—1 squadron of the Regiment A & E Troops under the command of Captain Newman marched from Hamilton Barracks this day and arrived at Newcastle on Tyne on 8th June, 1854.

30 June, 1854.—-The same squadron under the command of Captain Newman marched from Newcastle on Tyne this day and arrived at Sheffield on the 8th July, 1854.

28 June, 1854.—1 squadron of the Regiment C & F Troops marched from Piershill Barracks, Edinburgh, this day, and arrived at Nottingham Barracks on the 18th July, 1854, under the command of Captain Babington.

2 August, 1854.—One Squadron B & D Troops and Headquarters two divisions under the command of Major Hagart marched from Piershill Barracks, Edinburgh, on the 2nd and 3rd August, and arrived at Leeds Barracks on the 17th and 18th of the same month.

Inspected by Major-General Sir J. Thackwell, G.C.B. 24th August, 1854.

9 October, 1854.—One squadron (B & D Troops) with the band, under the command of Lieut.-Col. Hagart, proceeded by march route to Hull for the purpose of attending Her Majesty during her stay at that place, and returned to Leeds Barracks on the 18th of October. The Queen and the Prince Consort had been in residence at Balmoral and were returning south. Her Majesty and the Royal Party rested one night at Holyrood Palace and then came on to Hull where they occupied the Station Hotel. On Oct. 14th the Queen visited the Grimsby Docks in the Royal Yacht *Fairy* and returned to London by the Great Northern Railway that night.

8 November, 1854—One troop D and Headquarters , under

THE ROYAL VISIT TO HULL, 1854

the command of Lieut.-Col. Hagart, marched from Leeds this day, and arrived at York Barracks an the 9th Nov.

26 December, 1854.—One troop B, under the command of Captain Pedder, marched from Leeds this day and arrived at Nottingham Barracks on the 29th Dec., 1854.

20 February, 1855.—One troop A, under the command of Lieut. Brown, marched from York this day and arrived Sheffield Barracks on the 23rd.

3 July, 1855.—The Regiment was inspected by Major-General Arbuthnot.

28 August, 1855.—One troop C marched from Nottingham this day and arrived at York under the command of Captain Babington on Sept. 1st.

29 August, 1855.—One squadron(B & F Troops) marched from Nottingham this day under the command of Captain Pedder and arrived at Sheffield Barracks on the 30th August 1855.

In accordance with authority received from the War Office of this date authorising the addition of two troops to the Regiment, the following will be the establishment, viz. 8 troops, 1 colonel, 1 lieut.-colonel, 1 major, 8 captains, 8 lieutenants, 8 cornets, 1 paymaster, 1 adjutant, 1 riding master, 1 quartermaster, 1 surgeon, 2 assistant surgeons, 1 veterinary-surgeon, 1 reg. sergt. major, 8 troop sergt. majors, 1 paymaster sergeant, 1 armourer sergt., 1 saddler sergeant, 1 farrier sergt., 1 schoolmaster; to be appointed by the Secretary at War, 1 hospital sergt., 1 orderly room clerk, 31 sergeants, 31 corporals, 1 trumpet major, 8 trumpeters, 8 farriers, 600 privates. The two additional troops to be designated G & H.

6 November, 1855.—One troop H under the command of Cornet Seymour, marched from York this day, and arrived at Sheffield Barracks on Nov. 9th.

1 April, 1856.—A tunic substituted, in place of the pelisse and lace jacket.

2 July, 1856.—One squadron F & H troops under the command of Brevet-Major G. W. C. Jackson marched from Sheffield this day and arrived at Coventry on July 7th, received a fresh route and marched from Coventry on July 8th for the camp at Aldershot, where they arrived on July 15th.

7 July, 1856.—One squadron A & B troops under the com-

mand of Captain Pedder marched from Sheffield this day and arrived at Coventry on July 11th, received a fresh route and marched from Coventry on July 12th for the camp at Aldershot where they arrived on July 18th.

28 August, 1856.—One squadron D & E troops under the command of Major J. M. Hagart, marched from York this day and arrived at Northampton on Sept. 8th, received a fresh route and marched from Northampton on Sept. 9th for the camp at Aldershot where they arrived on Sept. 13th.

29 August, 1856.—One squadron C & G troops with the Headquarters of the Regiment under the command of Colonel C. Hagart. marched from York this day and arrived at Northampton on *Sept.* 9th, received a fresh route and marched from Northampton on the 11th Sept. for the camp at Aldershot, where they arrived on Sept 15th.

25 September, 1856.—Headquarters and 2 squadrons (E, F, G, H troops) marched from Aldershot to Guildford and went into billets there

A squadron (A & B troops) marched to Godalming on the same day and went into billets.

A squadron (C & D troops) also marched to Odiham and went into billets.

17 October, 1856.—G & H troops returned to Aldershot.

27 October, 1856.—E & F troops with Headquarters returned to Aldershot.

28 October, 1856.—C & D troops returned to Aldershot, as also did A & B troops.

31 October, 1856.—The Regiment was inspected by Major-General the Hon. Sir J.Y. Scarlett, K.C.B.

The Regiment having had its establishment augmented by two troops on August 29th 1855, on November 10th 1856 its establishment was reduced by two troops. The establishment of the Regiment therefore now became as it had been prior to that date. The total numbers were 470; there were 300 troop horses with 50 horses held supernumerary until they should be absorbed into the regular establishment. Two Captains, *viz.* D. P. Brown and J. Aytoun, were placed on half-pay, but the lieutenants and cornets 'exceeding the number of the establishment were ordered to be borne as supernumeraries of the corps until they shall fall into clear vacancies, or shall be otherwise provided for.'

30 April, 1856.—Another change took place.

In accordance with authority received from the Horse Guards, the authorized strength of the Regiment to be as follows, viz. 9 staff sergeants, 7 trumpeters, 6 farriers, 344 rank & file, 271 horses, although the effectives of the Regiment have been ordered to be reduced the Establishment is to remain unaltered.

30 May, 1857.—The Regiment was inspected by Major-General Sir James Y. Scarlett, K.C.B.

19 July, 1857.—500 of Sharp's breech loading carbines and 21 new patterned rifled pistols received from the Tower for the use of the Regiment and the old Victoria carbines and old pattern pistols given into store.

The 7th (Queen's Own) Hussars were now about to proceed to India for the first time. We shall deal with their services there and the stirring events of the Mutiny in the following chapters.

CHAPTER 4

The Indian Mutiny
1857

We have now reached the period which embraces the Indian Mutiny. during which the record of the Regiment is as full of interest for the reader as it is of honour to the Regiment. Before, however, we arrive at actual active service we must record such details of the voyage as have come down to us, and also give as many details of the arrangements which were made to bring the 7th (Queen's Own) Hussars up to its war strength.

On 6 August 1857, the Regiment having received orders for service in the East Indies, and in accordance with authority received from the War Office of this date, the following will be the establishment: viz., 10 troops (i.e. 9 troops abroad and 1 recruiting troop at home), 1 colonel, 2 lieut.-colonels, 2 majors, 10 captains, 10 lieutenants, 10 cornets, 1 paymaster, 1 adjutant, 1 riding-master, 1 quartermaster, 1 surgeon, 2 assistant surgeons, 1 veterinary surgeon, 1 regimental sergeant-major, 10 troop sergeant-majors, 1 quartermaster sergeant, 1 armourer sergeant, 1 saddler sergeant, 1 farrier sergeant, 1 schoolmaster (to be appointed by the Secretary of State for War), 1 hospital sergeant, 1 orderly room clerk, 40 sergeants, 1 trumpet major, 13 trumpeters, 40 corporals, 10 farriers, 626 privates—791 total numbers; 703 troop horses. The depôt troop to consist of 1 captain, 2 lieutenants, 1 troop sergeant-major, 8 sergeants, 8 corporals, 4 trumpeters, 20 privates; total number, 44. Apparently all the cornets went to India.

On 8 August the Regiment handed over its horses at Aldershot, and entrained for Canterbury.

Here it was made up to full strength by volunteers from the mentioned regiments: 1st Royal Dragoons, 20; 2nd Dragoons, 4; 3rd Light Dragoons, 1; 4th Dragoon Guards, 9; 4th Light Dragoons. 42; 5th Dragoon Guards, 21; 6th Dragoons, 17; 7th Dragoon Guards, 7; 8th Hussars, 2; 9th Lancers, 1; 10th Hussars, 9; 11th Hussars, 14; 13th Light Dragoons, 13; 16th Lancers, 18; 17th Lancers, 4; General Service, 1; total, 183.

On 14 August Captain the Hon. Ivo Fiennes, Lieutenant W. H. Seymour, Riding-Master J. Mould, and three sergeants proceeded by the steamship *Candia* from Southampton *en route* to India for the purpose of selecting horses for the Regiment in that country. The party arrived in India on 5 November.

On 27 August the service troops proceeded from Canterbury to Gravesend by rail and embarked on board the Clipper ship *Lightning*. Lieut.-Colonel James M. Hagart was in command.

Major Sir W. Russell, Bart., sailed in the same vessel, as did 4 captains, 11 subalterns, 5 staff officers, 43 sergeants, 10 trumpeters, 10 farriers, 33 corporals and 415 privates.

The *Lightning* sailed the same day. Colonel C. Hagart,Veterinary Surgeon J. Baker, and two sergeants were left at the depôt for the purpose of proceeding to India by the overland route to select horses for the Regiment.

The 7th Queen's Own Hussars arrived and cast anchor in the river Hoogly at Calcutta on 27 November. The Regiment disembarked on November 30 and encamped on the glacis of Fort William, Calcutta. One private died on the passage. An interesting record of the voyage of the *Lightning* has been kindly furnished by Colonel E. N. Pedder, late 13th Hussars. It takes the form a map on which the track of the vessel for each day and the distance sailed is accurately given. The voyage occupied eighty-eight days. At the time there were two officers of the name of Pedder in the 7th Hussars, Richard Newsham Pedder and Thomas Pedder—uncle and nephew. The map is the work of the former, at the time a lieutenant. He was father of Colonel E. Newsham Pedder, of the 13th Hussars. By the coarse marked out on the map we learn that on 2 October the *Lightning* was within sixty or seventy miles of the coast of Brazil, and for at least three days continued a path practically along the coast. The reason for this somewhat strange route was as follows. The vessel crossed the Equator at about 28° W. This was too far west. The south-east trade winds caught the ship and jammed it down close to the coast of South America, it being, as we see, at one time only from sixty to seventy miles distant there from. A comparison between this route and that which is now taken is not without interest.

We left the Regiment encamped on the glacis of Fort William. Here it remained till 5 December, on which and on two other dates it proceeded by train to Raneegunge. Two days later the Regiment marched by Bullock Dak to Allahabad in divisions; the First Division, consisting of A, B, and E Troops under the command of Captain T. Pedder, arrived at its destination on 19 December.

The Second Division, consisting of C, D, and F Troops, under the command of Captain W. D. Bushe, started on 8 December and arrived on 20 December.

The Third Division, consisting of G and H Troops with headquarters under the command of Lieut.-Colonel J. M. Hagart, started on 9 December and arrived on 21 December.

From the following telegram sent to the Governor-General of India in Council by the Commander-in-Chief and dated Cawnpore, 11 December 1857, 12 noon, we gather that the Regiment was then apparently at Benares, though this is not mentioned in their records; unless indeed it is to be translated as merely a change of destination already ordered from Benares to Allahabad. The telegram runs:

> The guns taken by General Grant will be in to-morrow. I shall be prepared to move forward in two or three days. Will your Lordship oblige me with any particular instructions you may have to give? I have desired the 7th Hussars to come up to Allahabad, to be formed under the personal superintendence of Brigadier Campbell, as their discipline would have suffered under the different authorities at Benares.

The Regiment remained at Allahabad until 17 January 1858, during which time the horses were received. The horses came from the Remount Depôt at Allahabad, and when handed over were untrained. By hard work, however, the Regiment got their mounts into order in a very short time, and on being inspected on 13 January by Brigadier W. Campbell they were reported 'fit for active service in the field.' Accordingly on 18 January they marched from

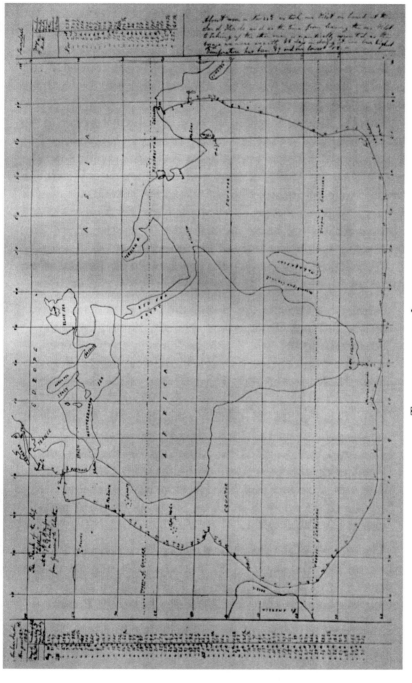

THE TRACK OF THE *LIGHTNING*

Allahabad under the command of Colonel C. Hagart for Cawnpore, where they duly arrived on 26 January.

It is somewhat strange to find that in the *Manuscript Regimental Record* there is not the faintest allusion to the outbreak of the Indian Mutiny—an event which, as we all know, began, continued, and ended in horrors and bloodshed tempered only with unexampled deeds of British heroism

That the Regiment proceeded upon actual active service in the field is mentioned in the entry dated 3 February 1858, and also that it was engaged with 'the enemy' a few days subsequently. Later the term 'infuriated fanatics' is employed. Once we are told that the 'enemy' were commanded by 'the Nana,' but beyond this clue, as far as the *Regimental Record* is concerned, the 7th (Queen's Own) Hussars, did we not know the real state of the case, might have been acting against an enemy of any nationality.

We will now now quote in the first instance the account of the campaign as given in the *Manuscript*, which begins abruptly thus:

February.—Three troops crossed the Ganges at Cawnpore on the 3rd and Headquarters with the remainder of the Regiment on the 4th. Employed keeping the road open, and escorting convoys between Cawnpore and Lucknow from the 4th to the 24th. The left wing of the Regiment was detached on an expedition under Major-General Sir J. Hope Grant on the 12th, and was engaged with the enemy at Meangunge on the 23rd when 5 men were wounded. One squadron marched for Alumbagh on the 24th and Headquarters and one squadron on the 25th. On the latter date two squadrons and Headquarters were present at the repulse of the enemy's attack on Sir James Outram's position at the Alumbagh.

The strength of the detachment engaged in repelling the fourth attack on the Alumbagh on 25 *February* 1858 was ninety-two, Colonel Hagart being in command. There were no casualties. To resume our quotation from the *Manuscript Regimental Record*:

March.—The Regiment was present during the whole of the operations of the siege and capture of Lucknow from the 1st to the 21st of March, under His Excellency The Commander in Chief (Sir Colin Campbell). On the 19th in a skirmish near the Moosabagh, Captain Slade and Lieutenant Wilkin were severely wounded, Cornet W. G. Bankes mortally and 2 men wounded. The latter Officer particularly distinguished himself when his Captain (Slade) was wounded, by gallantly leading on the troop, and thrice charging a body of infuriated fanatics, who had rushed on the guns employed in shelling a small mud fort, killing three of the enemy with his own hand, and receiving 11 wounds of which he afterwards died. He was awarded the V. C. for his gallantry on this occasion.

At Moosabagh camp on 21 March 1858, the strength of the Regiment is given as 350 sabres. Again quoting the *Manuscript Regimental Record*:

April.—Marched from the old cantonments Lucknow on the 11th, on an expedition under Major General Sir J. H. Grant; Engaged with the enemy at Baree on the 13th when one man was killed in a charge, and Lieutenant R. Topham and 6 men wounded. In Major General Grant's despatch of this affair he states:

The rapidity and steadiness with which the cavalry under Lieutenant Colonel Hagart was manoeuvred on my right and right rear is deserving of great praise.

Returned to Lucknow on the 27th and marched on the 28th on another expedition under Sir Jas. H. Grant towards Roy Bareilly.

May.—The Regiment was present at the affair of Cimsee or Sirsee on the 12th. One man wounded. In Sir J. H. Grant's despatch of this fair he states:

Our column was almost surrounded at one time, but the cavalry and artillery, the former commanded by Lieut Colonel Hagart, a very superior Officer, succeeded in clearing our right flank.

Returned to Lucknow on the 21st and marched from thence towards Poorwah on the 24th with a force under Major General Sir Hope Grant.

June.—Returned to Lucknow on the 11th June and marched

with a force under Sir Hope Grant for Nawabgunge, Bara Bunka, on the 12th. Engaged with the enemy at Nawabgunge on the 13th when Captain Fraser, Lieutenant Topham and Adjutant J Mould, and 15 men were wounded. 9 men died of *coup-de soleil*. In Sir Hope Grant's despatch of this engagement he states:

> The action on my proper right having again commenced with great vigour, I proceeded in that direction. On arriving I found two guns had come out on the open plain, and attacked Hodson's Horse with two guns of Major Carleton's Battery, which covered my rear. I immediately ordered up the other four guns under the command of Lieutenant Percival and two squadrons of H.M.s 7th Hussars under Major Sir W. Russell and opened grape upon the force within 3 or 400 yards with terrible effect, but the rebels made the most determined resistance and two men in the midst of a shower of grape brought forward two green standards which they placed in the ground beside their guns and rallied their men. Captain Atherley's two companies of the 3rd B. R. Brigade, at this moment advanced to the attack, which obliged the rebels to move off. The cavalry then got between them and the guns and the 7th Hussars led gallantly by Major Sir W. Russell, supported by Hodson's Horse under Major Daly C. B., swept through them twice, killing

THE RESIDENCY, LUCKNOW

every man. I must here mention the gallant conduct of two officers of the 7th Hussars, Captain Bushe and Captain Fraser. The latter I myself saw surrounded by the enemy, and fighting his way through them all, he was severely wounded in his hand.

July.—The Regiment remained encamped at Nawabgunge until the 21st July, when it marched for Fyzabad, where it arrived on the 29th.

August.—A wing of the Regiment marched for Sultanpore on the 9th with a force under Brigadier Horsford, C.B., where it arrived on the 13th. Headquarters with the remaining wing marched with a force under Sir J. H. Grant on the 19th and arrived at Sultanpore on the 22nd. In many places along the route the track led across cultivation and through marshes (the latter caused by heavy rains) where the gun wheels sank to the axle. The infantry were frequently obliged to wade through sloughs. On arrival it was found that the enemy amounting to 20,000 men with 15 guns, occupied a strong position and opposed the passage of the Goomtee. They had taken away or destroyed every boat, so that no bridge could be thrown across the river, which is upwards of 400 feet wide, and the right bank being in possession of the enemy for about 15 miles up and down the river, it was found impracticable to bring boats from a distance. Three small dinghies (dugouts?) were found concealed, and of these a good raft was constructed. Three more dinghies were brought from Biswa Nuddee, 9 miles distant, and three others were found sunk in the river. Of these two more rafts were constructed; the heavy guns having in the meantime been got into position to cover the operations and keep down the fire of the enemy.

The force commenced passing over by means of the rafts on the morning of the 25th. There was much difficulty experienced in swimming the horses across the river which is very deep and rapid. Only one horse of the 7th got drowned, altho' it took the Regiment two days to cross and all the force did not get over until late on the 27th. On the afternoon of the 28th the enemy came out in strong force and attacked our position, when they were repulsed and driven back, but

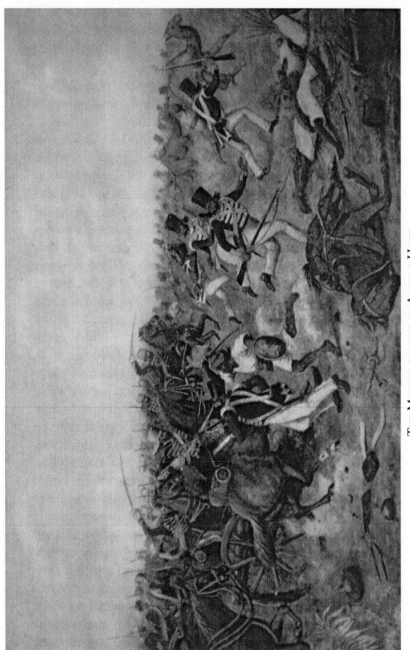

The Mutiny painted by Angelo Hayes

from the increasing darkness the troops were obliged to bivouac, and on advancing the following morning the cantonments (of the enemy of course) were found to be deserted.

Extract from a letter from the Adjutant-General of the Army to the Secretary to the Government of India, Military Department, with the Governor-General:

I am directed by His Excellency to beg you will draw the attention of the Right Honourable The Governor-General to the admirable manner in which Sir J. Hope Grant has conducted the operations of the last six weeks, and more particularly those for the passage of the Goomtee with the most imperfect means.

Extract from G. O. G. G. No.449 of 1858:

The Right Honourable The Governor-General desires to make known his high approbation of the military skill displayed by Sir Hope Grant during the series of operations which occupied six weeks and ended n the passage of the Goomtee. His Lordship also tenders to Brigadier Horsford, C.B., and to all the officers and men engaged in these operations his warmest acknowledgements for their gallantry and efficient services.

September.—One Squadron marched for Birtypoor and Silka on the 6th September, with a small force under Colonel Payne, H. M.'s 53rd, employed keeping the road open between Fyzabad and Sultanpore; a third troop marched to join them on the 28th.

October.—One troop marched on an expedition with a small force under Major Raikes, Madras Fusiliers, towards the Azimgurh District n the 7th. One troop front Silka to Headquarters on the 9th. Headquarters with a small force under Sir J. H. Grant marched towards the Azimgurh district on the 11th, a squadron being left at Sultanpore. Headquarters returned to Sultanpore on the 23rd, and the troop which was detached with Major Raikes' expedition on the 24th. This troop was engaged with the enemy in a skirmish at Shahpore on the 13th. The squadron which had been left at Sultanpore with a force under Brigadier Horsford, C.B., was engaged at the affair of Doudpore on the 20th, two Horse Artillery guns taken. The squadron at Silka rejoined Headquarters on the 25th, and the Regiment marched towards Amutha (?) on the 26th. Engaged in the Byswara campaign, under His Excellency the Commander-in-Chief until its completion on the 2nd December, including the pursuit of the enemy from Kandoo Nuddee on the 27th October, when the Regiment took two guns. The pursuit to Palee Ghât on the 28th, one gun taken.

November.—In attack on and pursuit of Benke Madho on the 29th, driving him across the Goomtee.

In Sir W. Russell's despatch of this pursuit he states "The 7th Hussars under Major F. W. Home were as usual perfectly steady."

December.—Arrived at Lucknow on the 3rd December, marched for Byram Ghât on the 5th, arrived on the 6th, marched for Fyzabad on the 8th, arrived on the 10th,— Crossed the Gogra on the 11th—Engaged in the Trans-Gogra campaign under His Excellency the Commander-in-Chief till its completion on the 31st December, including the attack on the enemy's position near Churda on the 26th, and pursuit in which the Regiment took 6 guns—The attack on the fort Mujeedia on the 27th—The final attack on the enemy's position near Banku on the 31st and pursuit into the River Raptee—Seven guns taken—Major Home and two men drowned and one man wounded.

In Lord Clyde's despatch of the Trans-Gogra Campaign he states:

On the 29th (December) the troops returned to Naupa-rah, made a forced march on the night of the 30th to the vicinity of Bankee where the enemy had loitered under the Nana. He was surprised and attacked with great vigour, driven through a jungle which he attempted to defend, and finally into and across the Raptee, the 7th Hussars entering that river with the fugitives. On this occasion the troops distinguished themselves, more particularly the 2nd Battalion R. Brigade under Colonel Hill and the 7th Hussars under Lieut.-Colonel Sir W. Russell. I have to deplore the loss of Major Home, 7th Hussars, who was drowned in the waters

Pursuit of the Rebels, December 29th, 1858

of the Raptee. He led the left wing of the Regiment. Captain Stisted who led the 1st Squadron was rescued with difficulty from a similar fate.

January, 1859.—The Regiment remained encamped on the banks of the Raptee during the whole of the month of January.

February.—Crossed the Raptee into Nepaul with a force under Brigadier Horsford, C.B. present at the attack on the enemy's position near Silka Ghât on the 9th. 14 guns taken. Remained in Nepaul until the 17th when the force recrossed the Raptee, and encamped at Sidhonia Ghât. The Regiment marched from Sidhonia Ghât on the 21st for Umballa, where it arrived on the 8th April after 47 days march. On the Regiment leaving Sidhonia Ghât the following order was issued to the Field Force. Extract from Field Force Orders by Brigadier Horsford, C.B., Commanding Field Force:

Camp
Sidhonia Ghât
21st February, 1859

For nearly a year, including trials entailed by a summer campaign, the 7th Hussars have been attached to this force, and during that time their behaviour has been such, as to secure them universal esteem. Brigadier Horsford begs Lieut-Colonel Sir William Russell Bart. will explain to his Regiment that he has thoroughly appreciated its state of discipline, and in wishing it the success it deserves, he feels satisfied that the just pride which has guided all ranks, not only in the face of the enemy, but also in camp, will ever cause them to retain the reputation they have so well earned.

This was signed by Major M. A. Dillon, Major of Brigade.

We will now attempt to amplify the above account, and shall first consider the events of February 1858. The Regiment formed part of the 2nd Brigade of the cavalry division which was commanded by Brigadier-General James Hope Grant, C.B (afterwards Sir James). The brigade was composed:

Brigadier W. Campbell, 2nd Dragoons Guards; Captain H. Forbes, 1st Light Cavalry, Major of Brigade (in these days Brigade Major). The regiments of the 2nd Brigade were: 2nd Dragoon Guards; 7th (Queen's Own) Hussars; Volunteer Cavalry; Detachment 1st Punjab Cavalry; Hodson's Horse.

About the middle of February it became known that bodies of the enemy were in motion near the fords or ghâts on the left bank of the Ganges, between Futeghur and Cawnpore; and that they were ripe for mischief. It became advisable to clear these rebels away, and for that purpose a moveable column was organised consisting of H.M. 34th, 38th and 53rd Regiments, squadrons of the 7th Hussars and 9th Lancer, squadrons of Hodson's Horse and Watson's Horse, a company of sappers and miners, and a few guns. The 34th, 38th, and 53rd Regiments belonged to the 2nd Infantry Division, and the 9th Lancers to the 1st Cavalry Brigade.

This column started from the main Lucknow Road at a point near Bunnee, and proceeded on a line inclining towards the Ganges at an angle which enabled it to sweep the rebels towards the west. Here they would be likely to be less mischievous than if they were near the river. The strength of Grant's column was as follows: cavalry, 636; infantry, 2240; artillery, 326; sappers (native), 44.

On 2 March 1858 the strength of the Regiment appears as follows: 30 officers; 392 non-commissioned officers and men; grand total, 422; horses, 413.

This telegram gives details of a reconnaissance:

7 March, 1858. 2.30 p.m.
From Brigadier-General F. H. Franks, C.B., commanding 4th Division, to the Deputy Adjutant-General of the Army, dated Headquarters camp Dilkoosha, 7 March 1858, 2.30 p.m.
I have the honour to report that at 9 a.m. this day an attack on our posts on the extreme left having been reported by a party of Hodson's Horse stationed there, I moved out with three battalions of Gurkhas and two of their guns, three guns of Lieutenant-Colonel Anderson's troop of Horse Artillery, the outlying picquet 7th Hussars, and Hodson's Horse, when the enemy (who were only seen at a distance) after a few rounds from the Horse Artillery and Gurkha guns retired into the city.
Those of the enemy seen were a few horse and several hundred infantry. They appeared to have no guns with them. The

troops returned to camp at 1 p.m.; no casualties took place, the rebels having only discharged a fire of matchlocks at too great a distance to do any harm.

The column was kept constantly on the move. Information was then received that a body of the enemy was strongly posted at Meangunge, a town situated between Lucknow and Futeghur.

The rebel force amounted to two thousand infantry in the town, three hundred cavalry outside, and five or six guns. Meangunge was surrounded by a stone wall fourteen feet high with numerous bastions and had three strong gates opening into the Lucknow, Cawnpore, and Rohilcund roads. At each gate the enemy had posted guns behind strong breastworks, and the breastworks themselves were covered by trees.

Grant reconnoitred the town carefully and found a weak point on the fourth side, where it was possible to bring two heavy guns to bear at the short range of between three hundred and four hundred yards, and at a spot where there was a postern-gate It will be observed that Grant had the superior force. He thus distributed his men. A part was told off to hold the Lucknow road, another was similarly employed on the Rohilcund road, while the remainder were posted behind a village which was conveniently near, with orders to await the result of a cannonade. Grant's guns opened fire, and in less than two hours a practicable breach in the wall was the result.

The 53rd were then ordered to advance to the assault. The regiment was divided. One wing under Colonel English after entering the breach worked through the left of the town, the other under Major Payne penetrated and turned to the right. The 53rd thus advanced through a labyrinth of narrow lanes, driving enemy before them at every yard. Meangunge was speedily captured and with it six guns. The enemy made vain efforts to escape by means of the three gates, but were most severely handled, nearly one thousand of their number being killed or captured. The British loss was very small, and is reported to have been 2 killed and 19 wounded. From another account we gather the following details:

Grant took up his command of 8 February at Oonao. His orders in the first instance were to keep the road open on both sides of Cawnpore. He was then ordered to make a rapid march to a small fortified place called Futtepore Churassie, where the Nana (Nana

Sahib) was supposed to have taken refuge. Futtepore Churassie was about twenty-five miles north of the Cawnpore road and on the banks of the Ganges. After two days' hard marching they reached the place, but found that the Nana had fled. However, two small field pieces were captured while a party of rebels were endeavouring to remove them. The fort here was destroyed and the buildings burnt.

The force then marched to Bangurmow and encamped under a large *tope* of trees. This town was large and well built, and the inhabitants professed loyalty.

On 21 February they marched to Sooltangunj *en route* for Meangunge, an old and moderate-sized town, with a rectangular loopholed wall round it. The enemy were prepared to resist. Two sowars of the advanced guard who approached too near the walls were fired upon, and one was shot.

On finding that the rebels meant fighting, Grant changed the direction of the columns from the Rohilcund road, along which they had been marching, to the left. Here a spot was discovered whence the wall could be breached, and two guns of position were placed there.

Turner's 9-pounder troop was posted a little further back to play upon the town and occupy the attention of the enemy. Four guns of Anderson's Troop, with the 7th (Queen's Own) Hussars, were sent along the Cawnpore road to check a threatened flank attack by a body of the enemy.

The other two guns, with the 34th Regiment and a troop of cavalry, were left on the Rohilcund road to cover the baggage. After an hour's cannonade a practicable breach was made and the 53rd were ordered to storm the town. In a few minutes they were pouring into the place 'like wildfire' and carrying everything before them. It appears that the rebels did not then know that the British had forced their way into the town.

Many were shot there and then, and many managed to rush out through the gates. Here they were at the mercy of the 7th Hussars, the 9th Lancers, and the Irregular Cavalry, who cut them down and ran them through without quarter; five hundred men killed and four hundred made prisoners.

We now come to the events at the Alumbagh, where Outram had been actively employed in repelling the energetic attacks of the enemy. As has been already mentioned, one squadron of the 7th marched thither on 24 February, and a second on the following day. On 25 February the last effort of the enemy against the Alumbagh took place. Reinforcements for Outram had been arriving for several days, and among these were the 7th Hussars (two squadrons), Hodson's Horse, the troop of Horse Artillery commanded by Remmington, and the 1st Bengal Fusiliers. This last attack took place under the immediate auspices of the Begum, who, mounted on an elephant, and accompanied by her prime minister and principal chiefs, witnessed the operations. A heavy cannonade of artillery began at 7 a.m. and lasted for an hour. Next a strong force was sent to threaten the British left, while main force proceeded along the right front of the British, but out of range.

This force, which has been estimated at between twenty and thirty thousand men, was divided; one half passed beyond the end of the British right, then turned to the right and took up a position in the Jellalabad, where it was sheltered by some convenient trees. The other half stopped short at the point beyond the extreme British right where the march of the first portion had been deflected to the right, and was prepared either to act in support of the attack or to maintain communications as occasion might need. With this portion of the force the Begum remained.

The force which had approached Jellalabad proceeded to shell the fort there. Against these Outram at once proceeded to act. Sending his right infantry brigade, four guns under Olpherts and four under Remmington, the 7th Hussars, detachments of Hodson's Horse and Graham's Horse, and also the military train, he started forth and cut off the advanced half of the enemy from the rest. Meanwhile he had detached the volunteers under Barrow and Wale's Horse to ride round and take them in the rear.

JELLALABAD FORT

He now found himself threatened by the Begum's force, who approached his left flank, and it appeared as if he would be attacked by some thousand of the enemy's cavalry besides infantry. The guns under Olpherts immediately opened on them and with effect. The Begum and her court fled incontinently. Remmington's four gun; galloped up and opened fire, while the 7th Hussars and Brasyer's Sikhs advanced as well. By this means the enemy in this part of the field was held in check.

Outram was then able to turn his attention to the other portion of the enemy, and after some fighting drove them back to shelter amid a clump of trees. By this time it was half-past two in the afternoon, and apparently the engagement was at an end. The enemy, however, determined on a last effort, and nearly three hours later made a strong attack on a village in the left front of Outram's position. They actually took possession of a clump of trees in front of it, but were not permitted to hold it long. Still, though everywhere held in check, the enemy did not give in, and throughout the night the struggle continued. With dawn, however, the enemy withdrew.

During the siege of Lucknow, which began on 2 March, the 7th Hussars were thus occupied. A detachment remained as a portion of the comparatively slender garrison which was left in the Alumbagh ; and took part in the events of 16 March, when, under the command of Brigadier Franklyn, an attack of the enemy was repelled, though not without severe fighting, as the engagement lasted from 9 a.m. till 1.30 p.m.

The remainder of the Regiment had proceeded with the cavalry of Sir Hope Grant, Lugard's division, and a strong force of artillery to circle round to the right from the rear of the Alumbagh, passing Jellalabad and going towards Dil Koosha. Here they took possession of the Dil Koosha Palace and the Mahomed Bagh. Batteries were erected which opened fire on the morrow with effect on the guns of the enemy and silenced them. Two bridges were at once begun to enable the troops to cross the Goomtee river, and the crossing was effected by daybreak on 6 March.

After crossing the Goomtee the column, under the command of Outram, moved to the north towards Chinhut. On reaching the Fyzabad road it turned to the left. At Ishmaelgunge the cavalry of the enemy were sighted and were attacked by the British cavalry.

LIEUTENANT–COLONEL HAGART

The rebels fled and were pursued, but meeting with broken and unfavourable ground in close proximity to the infantry of the enemy some losses occurred, mainly in the ranks of the Queen's Bays, which happened to be the leading regiment, their major (Percy Smyth) being killed.

The British halted at Ishmaelgunge during 7 and 8 March, during which time a heavy attack by the enemy was repelled on the first day, and the 9th Lancers with a battery of Horse Artillery were sent back across the river to the Commander-in-Chief, Sir Colin Campbell. Outram had now received his heavy guns, and hence the return of the Lancers and R H A. Two batteries were now erected, one against the rear of the Martinière, the other against Chukkur Kothi.

On the 9th the latter was captured. The combined column then advanced up the left bank of the Goomtee as far as the Badshah Bagh, which was seized and held. This enabled Outram to clear the enemy in flank and rear from both second lines of defences as far as the Kaiser Bagh Palace.

Meanwhile other battery had driven the rebels from the northern end of the first line of defences.

On 19 March Outram attacked the Musa Bagh. Sir Hope Grant, who was still on the left bank of the river Goomtee, was ordered to open fire on it and to prevent any of the enemy from crossing the river when driven out. Brigadier Campbell of the Queen's Bays was directed to take up a position on the left front of the Musa Bagh to cut off any of the rebels driven out by Outram. The enemy were duly expelled, but Brigadier Campbell failed to effect what he had to do and the enemy escaped. On this occasion a troop of the 7th Hussars was very actively employed.

It appears that near the position of Brigadier Campbell was a village with a small mud fort, the latter being held by the enemy. Under the command of Colonel James Hagart of the 7th a troop of the Regiment, two guns and a few men of the 78th were sent to dislodge them. The guns were brought up and opened fire. Two shells had exploded in the fort when out dashed about fifty of the enemy and made straight for the guns. The 7th were ordered by Colonel Hagart to charge, and did so. Three officers of the troop were wounded—Slade, Wilkin, and Bankes—the latter being mortally wounded, as has already been stated.

Colonel Hagart's gallant effort to save Bankes, who was being hacked at by the rebels as he lay disabled on the ground, so far succeeded in that that officer was rescued, though only to die of his wounds a few days later. Colonel Hagart's condition after the *mêlée* is thus related by Sir Hope Grant:'Everything about him bore traces of his gallant struggle. His saddle and his horse were slashed about both in front and behind, his martingale was divided, his sword-hilt dinted in, the pocket-handkerchief severed as clearly as with a razor, and a piece of the skin of his right hand cut away.' Sir Hope Grant recommended Colonel Hagart for the V. C., but Sir Colin Campbell declined to forward the application on the ground that his rank was too high. In the upshot every rebel was killed. After the combat Outram left the 2nd Punjab Infantry to occupy the Musa Bagh and returned to the positions he had occupied on the previous day.

Lucknow was now in the hands of the British. The city was then cleared of rebels by troops detailed for that purpose, the force under the Moulvi, which occupied a strongly fortified house, being the last to be reduced after a strenuous defence. His followers were pursued and cut up, but the Moulvi escaped.

On 23 March Sir Hope Grant, after a night march, attacked and defeated a body of the enemy stated to be four thousand strong at Kussi on the Fyzabad road and distant about twenty-five miles from Lucknow. The 7th Hussars were not, however, engaged on this occasion. The next proceedings of Sir Hope Grant were as follows. On 9 April he received orders to march at once to Barree, twenty-nine miles from Lucknow, where the Moulvi had collected a body of the enemy. Thence he was to march to Mohammadabad and along the river Gogra to reconnoitre a place called Bitaoli, where report had it that the Begum had taken up her quarters with a following of some six thousand rebels. He was afterwards to proceed to Ramnugger to cover the march of our allies, the Nepaulese troops, who were then on their return to their native land.

Sir Hope Grant started on the morning of the nth, and with his force went the 7th Hussars. His march lay along the Seetapore road, and his total strength amounted to about three thousand men with eighteen guns. The Moulvi by a clever piece of scouting obtained reliable information as to the column and prepared to profit

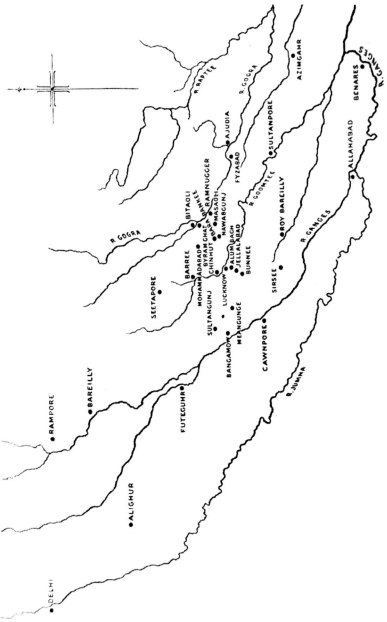

INDIA: OUDH & N.W. PROVINCES

by it. He occupied a village about four miles on the Barree side of the place where he had located the British camp. This village was covered by a stream in its front with high banks on the nearest side and the ground leading up being honeycombed. In this village the rebel infantry was posted, but the cavalry was sent to make a detour for the purpose of a flank attack.

Utterly un-suspicious, the British force marched at daybreak. The rebel cavalry was nearing a position in the rear which would have rendered the baggage—a train of some six thousand carts—an easy prey, when they chanced to catch sight of two British guns in the advance guard which were but slenderly escorted by Wale's Horse. They attacked the guns, wounded the officer in command of the escort, and were about to remove their trophies when a troop of the 7th Hussars, commanded by Captain Topham, appeared, and moreover appeared about to charge. The rebels awaited it not, but abandoning the captured guns fled at once. Two other attacks on the baggage in the rear were made, but without success, one of which was foiled by Captain Topham and his troop, and the other by a volley at very short range by two companies of the Bengal Fusiliers.

The rebel cavalry then retreated to the village. Grant pressed on his attack; the enemy, however, declined to offer any resistance, and evacuated the place. Grant then hurried on to Barree and then to Mohammadabad, where he arrived on the 15th. The next day he reached Ramnugger. He was now only six miles from Bitaoli, where it was rumoured that the Begum was established. Thither too he went with all speed, only to find the place evacuated. Finally he came up with the Nepaulese force under Jung Bahadur at Masaoli, a place half-way between Ramnugger and Nawabgunge. Thence he marched southwards to protect the road between Cawnpore and Lucknow, which was threatened at Onao.

Another account of the operations on the further side of the Goomtee may here be given, as it contains, though unofficial, a considerable amount of detail. It runs as follows:

The portion of the siege-plan connected with the left bank of the river had never been lost sight of during the preparatory operations on the right. While the cavalry, artillery, infantry and commissariat were busily engaged in camping near the Dil Koosha, the engineers were collecting the casks, fascines of fagots, ropes and timbers necessary for forming a bridge, or rather two bridges, across the Goomtee, at some point below where the enemy were in greatest force. The spot selected was near headquarters at Bibiapore, where the river was forty yards wide. The enemy, uneasy at the proceedings of the engineers, gradually assembled in considerable numbers on the opposite bank, but as the British brought up runs to oppose them, the engineering works proceeded without molestation. These bridges exemplified some of the contrivances which military commanders are accustomed to adopt, in the course of their onerous duties.

The groundwork of each was a collection of empty beer-casks, lashed by ropes to timber cross-pieces, and floated off one by one to their positions; a firm roadway of planking was afterwards fixed on the top of the whole range from end to end. Firm indeed must the construction necessarily have been, for troopers on their horses, heavy guns and mortars, ammunition-wagons and commissariat carts, all would have to pass over these bridges, secure so far as possible from accident to man or beast.

Outram crossed the river on the 6th in command, as we know, of a strong force. His plan of campaign was to advance up the left bank of the Goomtee; while the troops in position at Dil Koosha were to remain at rest until it was apparent that the first line of the enemy's works, of the rampart, running along the canal and abutting on the Goomtee, had been turned. The bridges stood the strain admirably, and the entire force passed over in safety. A little fighting took place, notably in front of the Chukkur Walla Kothi or Yellow House, a circular building on the left bank of the river. Meanwhile Sir Colin Campbell remained on the defensive near the Dil Koosha. The enemy's guns at the Martinière were active, but their practice was bad. It is stated that a Lieutenant Patrick Stewart was most energetic in establishing the electric telegraph, and that it was carried from camp to camp so that Calcutta, Allahabad, Cawnpore, the Alum Bagh and other places could be immediately linked up. Nor was this all. Wires to Outram across the river were established, and one was drawn through the window of the Dil Koosha itself. Wherever Sir Colin went, there the gallant and energetic Stewart was to be found with his poles, batteries, and wires. We will again quote:

On the 7th, Sir James Outram, while making his arrangements on the opposite side of the river, was attacked in great force by the enemy. On the preceding day, he had baffled them in all their attempts, with a loss of only two killed and ten wounded; and he was not now likely to be seriously affected even by four or five times his number. The enemy occupied the race-course stand with infantry, and bodies of cavalry galloped up to the same spot with the intention of disturbing Outram's camp. He resisted all the attacks, chased them to a distance with his cavalry, and maintained his advantageous camping-ground.

Russell, the celebrated war correspondent, watched the proceedings both days from the roof of the Dil Koosha. He describes the weather thus:

The wind was all but intolerable—very hot and very high, and surcharged with dust. I had a little camp-table and a chair placed on the top of the building, and tried to write; but the heat and dust were intolerable. I tried to look out, but the glasses were filled with dust; a fog would be just as good a medium.

The road from Fyzabad and from the cantonment passed near Outram's camp. On the 8th Sir Colin Campbell made a reconnaissance, with the result that Outram was ordered to arrange his batteries during the night and on the following day to attack the enemy's position, the key to which was the Chukkur Walla Kothi. We will again quote:

On the morning of the 9th, accordingly, Sir James made the attack with excellent effect, the enemy being driven out at all points, and the Yellow House seized. He advanced his whole force for some distance through ground affording excellent cover for the enemy. He was by that means enabled to bring his right wing forward to occupy the Fyzabad road, which he crossed by a bridge over a nullah, and to plant his batteries for the purpose of enfilading the works upon the canal.

The operations were most successful; there was much skirmishing and a most obstinate fight at the Yellow House, where a few fanatics had shut themselves in and maintained a most obstinate resistance. They were at length driven out or slain, and the occupation of the villages of Jeamoor and Jijowly followed. Next came an advance to the Padishah Bagh or King's Garden opposite the Fureed Buksh palace, after which Outram's guns opened a fire which enfiladed the lines of the Kaiser Bagh defences. Sir Colin Campbell now stormed and took the Martinière. Here the enemy declined a hand-to-hand contest and escaped from the walls and trenches. The British only employed the bayonet.

On the 10th, while Outram was engaged in strengthening the position which he had taken up, he sent Hope Grant with the cavalry of the division to patrol over the whole country between the left bank of the Goomtee and the old cantonment.

Much fighting took place on Colin Campbell's side of the river on this day. On the 11th Jung Bahadur with his Nepaulese arrived, tardily, it is true, but it may be noted that they had already rendered service in the Goruckpore and Jounpore districts, and had then advanced into Oude to assist in the operations against Lucknow. The battle was everywhere going on, yet time was found amid a crash to hold a ceremonial durbar. In front of Sir Colin's mess-tent he and the Nepaulese commander met, though with customary want of punctuality the Oriental was not there to time. During the ceremony news arrived that 'the Begum Kothee is taken.' The ceremony was hastily broken off. On the 13th the Nepaulese moved dose to the canal. Next day Jung Bahadur was requested to cross the canal and attack the suburbs to the left of Banks' house. Here the Nepaulese were 'most advantageously employed for several days' in covering the British left.

Outram meanwhile was constantly fighting. On the 15th two bodies of cavalry were sent out, one under Walpole, along the Sundeela road, and the other under Hope Grant on that which led to Seetapore. Gradually the British force meanwhile was working its way across the city, and by the 17th Sir Colin Campbell was virtually master of the situation.

On this day two English ladies. Mrs. Orr and Miss Jackson, were delivered from the hands of the enemy. The 18th was comparatively quiet. On the 19th a combined attack on the Musa Bagh was organised. The enemy fought not but fled, and unfortunately escaped slaughter or capture to a great extent.

Lucknow was now all but taken. Only the Moulvi remained: this worthy had shut himself up in a stronghold in the heart of the city. From this he was driven; his following was pursued for six miles and many were slain, but the Moulvi himself escaped to fight another day. During this period Hope Grant's cavalry had cut up a few hundred fugitive rebels in one place and intercepted more in another.

In *May* 1858, to make a short digression, it is interesting to note that white clothing was ordered to be discontinued in the European regiments of the Honourable East India Company's Army. Its place was taken by *khakee* or *carkey*—the khaki of these latter days. We need not do more than mention the fact. It is clear, however, that in the *Queen's Army* the regulation dress was found unsuitable for campaigning and extremely irksome to the troops. We read that—

Except the Highlanders—and when they left Lucknow they were panting for their summer clothes, and had sent officers to Cawnpore to hurry them—not a corps that I have seen sport a morsel of pink or show a fragment of English scarlet. The Highlanders wear eccentric shades of grey linen over their bonnets—the kilt is discarded, or worn out in some regiments; and flies, mosquitoes, and the sun are fast rendering it impossible in the others. Already many officers who can get trews have discarded the ponderous folds of woollen stuff tucked into massive wads over the hips, and have provided some defence against the baking of their calves by day, and have sought to protect their persons against the assaults of innumerable entomological enemies by night. The artillery had been furnished with excellent head-covers and good frocks of light stuff. . . . The 7th Hussars, the Military Train, have vestiary idiosyncrasies of their own; but there is some sort of uniformity among the men.

A good deal more follows, descriptive of the vagaries in head-gear—feathers, helmets, shooting-jackets, and Wellington boots. The writer continues:

The peculiarity of carkey is that the dyer seems to be unable to match it in any two pieces, and that it exhibits endless varieties of shade, varying with every washing, so that the effect is rather various than pleasing on the march or on the parade ground.

The writer might also have instanced the Blue Caps of the regiment now known as the 1st Battalion Royal Dublin Fusiliers as a specimen of non-regulation headgear which became historic. After the capture of Lucknow various military works were undertaken to place that important city in such a state of defence as to render it safe from attacks within and without; for it must be remembered that great though the slaughter of the rebels had been there, yet so large a number had escaped that the greatest precautions had need to be taken against renewed attacks.

Sir Hope Grant, to whom the conduct of military affairs in Oude had been entrusted, now undertook an expedition against a body of rebels who were rumoured to number from seventeen to eighteen thousand men, and to have crossed the River Gogra and taken up a position at Ramnugger Dhumaree. It was also affirmed that Madhoo Singh at the head of five thousand rebels was at Goosaengunge, Benhi Madhoo with a smaller force in the Poorwah district, and Dunkha Shah with a larger army at Chinhut. Probably these numbers were exaggerated; still it tot prudent to leave the north-eastern portion of Oude unprotected, and hence Grant organised his movable column and proceeded towards Fyzabad.

A little before midnight on June 12 he marched from Lucknow to Chinhut, and thence to Nawabgunge on the Fyzabad road. His force consisted of the 2nd and 3rd battalions of the Rifle Brigade, the 5th Punjaub Rifles, a detachment of engineers and sappers, the 7th Hussars, 2 squadrons of the 2nd Dragoon Guards, Hodson's Horse, a squadron of the 1st Sikh cavalry, a troop of mounted police, a troop of Horse Artillery, and two light field-batteries. Leaving a garrison column at Chinhut, under Colonel Purnell, and entrusting the same officer with temporary charge of the baggage and supplies belonging to the column, Sir Hope resumed his march during the night towards Nawabgunge, where 16,000 rebels had assembled, with several guns. By daylight on the following morning he crossed the Beti Nuddee at Quadrigunje, by means of a ford. He had purposely adopted this route instead of advancing to the bridge on the Fyzabad road, in order that, after crossing the nullah, he might get between the enemy and a large jungle. As a strong force of rebels defended the ford, a sharp artillery-fire, kept up by Mackinson's horse-artillery and Johnson's battery, was necessary to effect this passage. Having surmounted this obstacle, Sir Hope, approaching nearer to Nawabgunge. got into the jungle district. Here the rebels made an attempt to surround him on all sides and pick off his men by repeated volleys of musketry. The General speedily changed the aspect of affairs. He sent a troop of horse-artillery to the front; Johnson's battery and two squadrons of horse were sent to defend the left, while a larger body confronted the enemy on the right, where the enemy evidently expected to find and to capture Sir Hope's baggage.

The struggle was very fierce, and the slaughter of the rebels considerable; the enemy, fanatical as well as numerous, gave exercise for all Grant's boldness and sagacity in contending with them. The victory was complete, and yet it was indefinite; for the rebels, as usual, escaped, to renew their mischief at some other time and place. Nearly six hundred of their number were slain; the wounded were much more numerous. Hope Grant's list of killed and wounded numbered about a hundred. Many of the rebels were Ghazees or Mahommedan fanatics, far more difficult to deal with than the mutinied sepoys. Adverting to some of the operations on the right flank, Grant said in his despatch:

> On arriving at this point, I found that a large number of Ghazees, with two guns, had come out on the open plain, and attacked Hodson's Horse. I immediately ordered up the other four guns under Lieutenant Percival, and two squadrons of the 7th Hussars under Major Sir W. Russell, and opened grape upon them within three or four hundred yards with terrible effect. But the fanatics made the most determined resistance; and two men amid a shower of grape brought forward two green standards, which they planted in the ground beside their guns, and rallied their men. Captain Atherley's two companies of the 3rd Battalion Rifle Brigade at this moment advanced to the attack, which obliged the rebels to move off. The cavalry then got between them and

the guns; and the 7th Hussars, led gallantly by Sir W. Russell, supported by Hodson's Horse under Major Daly, swept through them, killing every man.

The main body of rebels succeeded in escaping from Nawabgunge after the battle. They fled chiefly to Ramnugger and Mahadeo on the banks of the Gogra, and to Bhilowlie at the junction of that river with the Chowka—with the apparent and probable intention of throwing up earthworks for the defence of those positions.

We will now add a few details to the account already given of the action at Nawabgunge.

The rebels had begun to collect again around Nawabgunge, a place situated about eighteen miles from Lucknow. There Sir Hope Grant determined to attack. He accordingly marched for Chinhut, where he learnt that the enemy were in great force at Nawabgunge. and moreover that their position was a very strong one. They had camped upon a large plateau which was surrounded on three sides by a stream, over which, at a little distance from the town, there was a good bridge. The fourth side was jungle.

Grant determined to turn the right of the enemy and to thrust himself between them and the jungle. Two miles up the stream there existed a platform bridge which he purposed to cross. His column started at night and had a distance of twelve miles to traverse. It was hoped that the enemy would be surprised. All baggage was left in the rear, and the force marched with the lightest possible equipment. The night was very dark when a start was made at 11 p.m., so much so that it was very difficult to find the way across the level open country for at least six miles. A guide who proved to be trustworthy was, however, luckily obtained. The march was successfully accomplished, though some of the men were unfortunately victims to heat apoplexy. The bridge was reached half an hour before dawn, and the column rested and fed. At daylight the men fell in. The enemy had two or three guns which commanded the bridge, but they were posted at too great a distance to be effective. They opened fire but were soon silenced by a battery of nine-pounders which Grant sent forward for that purpose, one gun being dismounted.

The stream was then crossed by four companies of rifles, a troop of horse-artillery, and some cavalry. These having effected a lodgement on the other bank, the main body of the column followed them. It was then found that the British fronted the very centre of the enemy's position; also that the surprise was accomplished and that the rebels were unable to concentrate. As a matter of fact the force of the rebels at Nawabgunge was composed of four separate bodies, each under its own leader, and each acting independently.

Great gallantry was shown by a body of fine daring Zemindaree men who got round the British rear with two guns and attacked in the open. These were the men who attacked Hodson's force and two of whose number planted the green standards by the guns which we have already mentioned. After they had attacked Hodson's Horse with some success and had menaced the two guns with that regiment, Grant ordered up the 7th Hussars and the other four guns belonging to the battery. These four guns were posted within five hundred yards of the enemy and opened with grape 'which mowed them down with terrible effect, like thistles before the scythe.' Their chief, the man who ordered the two green standards to be planted near the guns, is described as 'a big fellow with a *goître* on his neck.' Two squadrons of the 7th Hussars under Sir William Russell and as many companies of the rifles were now sent forward and forced the enemy to retire; but, though forced back, the rebels were still undaunted, and with waving swords, spears and abusive language, called on the British to 'come on.'

The 7th Hussars charged twice through them, and cut them up to a man. Round the guns alone lay one hundred and twenty-five dead.

The action lasted for three hours. Six of the rebel guns were taken and six hundred rebels slain. The British loss was sixty-seven killed and wounded; thirty-three deaths from sunstroke and two hundred and fifty more men stricken down and obliged to be sent into hospital. It is stated that 'men fell asleep in their tents and never awoke,' heat apoplexy being the cause of this excessive loss of life. Though six guns were captured, the enemy were possessed of many more, and these they succeeded in removing. Each of the four bodies of the enemy retreated in a different direction, and as it was impossible for Grant to follow them up with the small remaining force which was at his disposal they succeeded in making good their escape.

Grant himself returned to Lucknow, his force remaining at Nawabgunge. His next expedition was to march to the relief of the somewhat notorious Rajah Maun Singh. This worthy was both powerful and wealthy. He had been a rebel, but having deserted that cause, probably because he found it a fallen one, or had come to the conclusion that it would ultimately fail, was now professedly loyal to John Company.

The consequence was that he found himself shut up in a large mud fort defended by thick walls and a broad ditch, and besieged by twenty thousand rebels with twenty guns.

Starting from Nawabgunge on 22 July Grant's column proceeded for some eight miles along the Fyzabad road. Here intelligence was received that one thousand two hundred rebels were posted in a village twelve miles to the south-east. To clear these out Grant detached two hundred of the 7th Hussars, the same number of Hodson's Horse, a troop of Horse Artillery under Major Yates, and the 5th Punjab Infantry. Colonel Hagart of the 7th was in command of the whole. Colonel Hagart started at night hoping to surprise the rebels, but found them already departed. He rejoined Grant at Derriabad on 24 July. Maun Singh had meanwhile been sending messages imploring relief; but on the 26th his tone changed, and he informed Grant that the enemy was fast disappearing and that unless the British made haste they would all escape. The British did make haste, but found, when but a day's journey from their goal, that not a rebel was left for them to fight. Some had joined the Begum across the Gogra, and eight thousand in two parties of four thousand each had made their way to Sultanpore. Grant reached Fyzabad on 29 July. Here he halted for an hour and then marched to the Ghât of Ajudia, four miles down the river Gogra.

Here they caught several boats conveying fugitives across. Grant's guns opened on them and sank all but one; the crews, however, jumped overboard and swam ashore down the river. The boat that escaped was not hit, nor was it deserted.

On 2 August Grant, accompanied by two hundred cavalry, paid a return visit of ceremony to Maun Singh at his fort. Grant was now ordered to drive the enemy out of Sultanpore and to occupy that town. He therefore organised a detachment under the command of Brigadier Horsford, which marched on 7 August. Horsford arrived within three and a half miles of the place by 12 August. Sultanpore is a town situated on the banks of the Goomtee river. The enemy were found to be in force, and had guns posted to defend the passage.

Horsford's force was a small one. Rumour had it that the enemy numbered fourteen thousand men with fifteen guns. On the approach of the British to the river the enemy retired. Still Horsford did not feel justified in crossing, and halted, reported to Grant and awaited orders. The Commander-in-Chief had also heard the strength of the enemy at Sultanpore, and sent a telegraphic message to Grant to reinforce Horsford. This reinforcement Grant accompanied. He started on the morning of 19 August with all his guns of position and the whole of his available force—a force which had been strengthened by the arrival of the 53rd Regiment. He reached Sultanpore on the 22nd. The town lay in a bend of the river. Horsford had already selected a spot where he proposed to cross, and his selection met with Grant's approval. It was well sheltered from the fire of the enemy, and suitable in other respects. The only trouble was that the river could not be forded, and no boats were available. The enemy held the command of the river for fifteen miles both ways. Search for boats had hitherto proved in vain, till at length three 'small very rotten canoes, hollowed out of trees' were discovered. These boats are elsewhere called dingheys, but dug-outs would probably be the more correct designation.

Two Engineer officers, Captain Scott and Lieutenant Rainsforth, with a party of Madras sappers, were soon hard at work converting these crazy craft into a serviceable raft. By good luck six more were discovered, of which three had been sunk in a creek in the river and the rest were moored near, the finder being the Deputy-Commissioner Captain Reid. Out of these two more rafts were constructed, and with this slender power of flotation Grant's force embarked. The crossing was successfully accomplished, though one raft broke under the weight of a second nine-pounder gun, a canoe giving way and causing the whole to capsize. The guns were therefore dismounted and sent across singly. The cavalry swam the river, and with the loss of only two horses Grant's column found itself on the other side of the stream. The transit, however, occupied two days. Meanwhile the advance body, the Madras Fusiliers (*Blue Caps*) and the 5th Punjab Infantry had taken up a strong position.

The Rifles were then sent across in support, at which time the enemy's guns had opened fire. On the evening of the 28th the rebels made an attack which was repelled, but owing to the darkness it was not possible to pursue. On 29 August the enemy was found to have evacuated their position.

On 23 October Grant, who had again been absent—this time to be invested as K.C.B.—returned to Sultanpore and again started out on the warpath. This time his objective was Pandoo Nuddee, where four thousand rebels with two batteries of guns were in position, commanding the bridge. The rebels, however, did not wait to be attacked, but bolted in terror into the jungle behind their position. The 7th Hussars and some Horse Artillery were sent in pursuit, and after a long chase—it is said of some thirty miles— captured two guns, one of which was a brass twenty four-pounder. Very few of the enemy, however, were cut up.

We now reach the episode of the crossing of the river Raptee, which has already been touched upon.

It will be admitted that the fording of an Indian river is an operation which is not to be undertaken by a couple of regiments of cavalry at a fairly high speed, and without some knowledge of the conditions under which the crossing is to be effected. That is to say, it is well to know whether the bottom is unencumbered by rocks, whether the current is swift, or whether there are holes. In this case, however, it would appear that not only was the attempt made to ford an un-surveyed stream and at speed, but it was discovered too late that rocks, quicksands, trunks of trees, and holes abounded.

That the charge was gallantly made it is impossible to deny, but that it ever ought to have been made under the conditions which obtained at the spot cannot be reasonably affirmed. Hence it was that the Regiment had to morn the loss of the major and more than one of their comrades-in-arms.

Beneath a large picture in the mess which represents the incident the following account has been placed:

On 29th December, 1838, news was received that the Nana with his army was within 25 miles of our position on the borders of Napaul. That evening the column was set in motion, and at 5 a.m. arrived at a village about a mile from the enemy. The cavalry (7th Hussars and 1st Punjaub Cavalry) and R. H. A., under Sir William Russell were ordered to advance and drive the enemy from their position. They formed echelons of squadron on each flank with the guns in the centre, and the whole advanced over the plain at a gallop. The enemy could not stand this rapid attack, and abandoning their most advanced guns, fled, and were pursued to a long belt of jungle which stretched for miles, and were not dislodged until the infantry came up, driving them from their position and forcing them through the wood and over another plain which stretched down to the river Raptee about six miles off. The plain was covered with the army of the Nana and the largest body appeared to be making towards the lower ford to the right. As soon as the cavalry and artillery had defiled through the jungle, they were launched in pursuit to the right, but their career was stopped by a very wide and difficult *nullah*, which the cavalry got over, but which stopped the artillery. Sir William Russell led the 1st Punjaub cavalry supported by the 1st and 2nd squadrons, 7th Hussars, and charged to the right, driving the enemy into the river. Finding it impossible to force the ford in the face of the fire from a battery of heavy guns, Sir William Russell wheeled to the left, and running the gauntlet through a hot fire from the enemy's guns and musketry, galloped along the plain towards the upper ford on the left, to which the 3rd and 4th squadrons, 7th Hussars, under Major F. W. Home, had been previously sent.

The above picture represents the 3rd and 4th squadrons, 7th Hussars, rapidly closing in on the fugitives, who made for the ford, which was interspersed with rocks, quicksands, trunks of trees, etc. The speed became tremendous as they neared the enemy. As the word *charge* was given, a cheer rose from the ranks, and they closed with a shock—men and horses rolled together into the river, which, running like a mill stream, was alive with rebels, trying to escape. A scene of confusion ensued, better imagined than described. Lieut. Stewart saw a huge *sowar*, whose horse had fallen under him in the river, standing at bay with his *talwar* over his head, ready to strike. Stewart dropped his sword, by the sling and drawing

The charge at the Raptee

his revolver, shot him. Major Home, who led the leading squadron most gallantly, was last seen in the river, engaged with two *sowars*. His body was found in the water two days after by some native divers, under the trunk of a tree, with a dead sowar grasped in each hand. Captain Stisted was rolled over in the mêlée and nearly lost his life, his horse was carried down the stream and drowned. Sir William Russell, having galloped ahead to overtake these squadrons before they charged, came up at the moment they reached the river. He halted the squadrons as soon as possible and got the men together to prevent further loss of life which was now useless. Capt. Stisted and three men of the 7th Hussars whose horses had been drowned, were standing on a small sand bank in the middle of the river. None of them could swim, and as the river was running like a sluice, they had much difficulty in keeping their footing, and were in great danger of being drowned. Major C. C. Fraser (afterwards General Sir Chas. Fraser, K.C.B.), though at the time being partially disabled from wounds, volunteered to swim to their rescue and succeeded in saving them all after considerable difficulty, and

under a sharp musketry fire from the enemy on the opposite bank. He received the Victoria Cross for this gallant action. The figures, commencing from the left, are:
Capt. Thos. H. Stisted, his horse rearing over with him in the river; Lieut. R. D. Stewart, firing his pistol at the sowar; Lieut. H. J. Wilkin, his sword raised, leading on a troop; Lieut.-Col. Sir William Russell, in right corner with his left hand uplifted, halting the squadron with his usual "Steady, men, steady."

It is curious to note that *Malleson* (vol. 3. p. 295); in mentioning the charge into the Raptee by the 7th Hussars, miscalls Major Horne, stating his name to be Home. Regarding the recovery of the bodies of Major Horne and the two privates who were drowned, we read in the *Life of Sir James Hope Grant* that:

After some search the bodies were drawn out of a deep hole, Horne with a fast grip of two of the enemy, and the two privates each clutching a sowar. This was probably the result of the death struggle.

CHAPTER 5

The North West Frontier
1859

The record of the eleven years which elapsed after the suppression of the Indian Mutiny, before the Regiment was again at home, is not except for one event of great interest.

The first station occupied by the Regiment after hostilities had ceased was Umballa, and we find them there in January 1859. On 20 January one squadron (A and B Troops), under the command of Captain F. Garforth, marched thence to act as escort to the Governor-General (Viscount Canning), and returned to their cantonments on 14 April. We are not, however, told their destination.

The next entry is dated 30 July, 1860, there being an interval of no less than eighteen months without a single fact being recorded in the Regimental Manuscript. On that date we learn that 'In accordance with instructions received from the War Office' the establishment of the Regiment was to be altered, and that it was in future to consist of nine troops abroad and one at home. The strength of the Regiment was then fixed as follows: 1 colonel, 2 lieutenant-colonels, 2 majors, 10 captains, 10 lieutenants, 10 cornets, 1 paymaster, 1 adjutant, 1 riding-master, 1 quartermaster, 1 surgeon, 2 assistant surgeons, 1 veterinary surgeon, 1 regimental sergeant-major, 10 troop sergeants-major, 1 quartermaster-sergeant, 1 paymaster-sergeant, 1 armourer-sergeant, 1 saddler-sergeant, 1 farrier-major, 10 farriers, 1 hospital-sergeant, 1 orderly-room clerk, 40 sergeants, 1 trumpet-major, 13 trumpeters, 40 corporals, 10 shoeing smiths, 616 privates; total, 791; the number of troop horses being 703.

On 16 November, 1860, the Regiment was inspected by Major-General Sir R. Garrett, K.C.B. In 1861, on 23 February, his Excellency the Commander-in-Chief, Sir Hugh Rose, inspected the Regiment and was pleased to express himself highly satisfied with their efficiency; he observed that 'they had made the best cavalry field-day that he had seen.' The 7th Hussars were also inspected by Sir R Garrett, K.C.B., on 22 April.

The establishment of the Regiment was now again altered. We read under date 20 June, 1861, that it was in future to consist of the following numbers: Eight troops abroad and one at home. The strength was as before, but with these alterations: 1 lieutenant-colonel, 9 captains, 9 lieutenants, 9 cornets, 9 troop sergeants-major, 8 farriers, 32 sergeants, 12 trumpeters, 548 privates. The total strength was now six hundred and ninety-four and the horses numbered six hundred and seventeen. On 28 March, 1862, the Regiment was inspected by Brigadier-General Renny, C.B. Under the date 13 December, 1862, we read:

The Regiment under the command of Lieut.-Colonel A. Scudamore, C.B., marched this day from Umballa *en route* to Peshawur. at which station it arrived on the 6th of February 1863. The Regiment had remained at Umballa for 3 years and eight months.

According to which statement the 7th must have arrived at Umballa in April 1859. On 6 January 1863 the establishment of the Regiment was for the third time changed since July 1860. It now stood thus: Seven service troops abroad and one at home. Everything remained as before, with these exceptions: there were now 8 captains, 8 lieutenants and 8 cornets, 8 troop sergeants-major, 7 farriers, 29 sergeants, 36 corporals and 483 privates; total, 620. The number of horses is not mentioned. Extract from General Order:

Horse Guards
3rd September 1863
The Queen in commemoration of the services of the undermentioned in regiment in restoring order in Her Majesty's Indian Dominions, is graciously pleased to command that the word *Lucknow* be borne on their standards, and appointments:
For capture of Lucknow. 7th Queen's Own Hussars.
(Signed) *A. H. Horsford*
Deputy Adjutant-General

The monotony of cantonment life in 1863 and 1864 was pleasantly broken by two little frontier affairs. On 6 December 1863 E Troop, under command of Lieutenant Holmes, marched to Fort Ghub Kudder, one of the Peshawur outposts, distant about eighteen miles from the cantonments.

E Troop escorted half a battery of Horse Artillery, and the next day escorted these guns in an affair between the troops at the outpost and about two thousand of the Mohmand tribe, who appeared in force at the foot of the hills in front of the fort. On the 22nd B Troop and on the 27th G Troop joined the Doaba Field Force, the three troops B, E, and G being now under the command of Brevet-Major T. H. Stisted.

On 1 January 1864 the enemy made a demonstration, and on the afternoon of 2 January about five thousand of the different hill tribes advanced to within a short distance of the fort. The British force consisted of half a battery of Horse Artillery, three troops of the 7th Queen's Own Hussars, the 3rd Battalion Rifle Brigade, as well as some native cavalry and infantry. The force was under the command of Colonel McDonald of the Rifle Brigade. The British turned out to oppose the enemy. The affair began by the advance of a squadron of the 7th Hussars, who trotted to the front and charged the more advanced bodies of the enemy by separate troops, after which the three troops formed in two squadrons charged in line, and considering the unfavourable nature of the ground did great execution. The loss of the Regiment on this occasion was three men killed and seven wounded; three horses were killed, ten wounded, and one missing.

Colonel McDonald expressed his approbation of the gallant conduct of the three troops, and subsequently at a parade of the whole of the force his Excellency the Commander-in-Chief thanked the officers and men of the 7th Hussars for their behaviour in the affair of the 2nd, alluding particularly to their well-conceived and spirited charges. The three troops returned to headquarters on 16 January.

We are glad to be able to give rather more information with regard to this frontier affair. It appears that the Mohmand chiefs had given in their submission some time previously and quiet reigned—as far as a state of peace ever really did exist on the frontier—till 1863. At that time emissaries from the Akhund of Swat were sent over the hills bordering on the Peshawur valley, but only succeeded in stirring up the Mohmands. The Sultan, Mohammad Khan, who was the son of Saadhat Khan, owned the religious supremacy of the Akhund of Swat, and also hated the British—as indeed all did. He was known to be the murderer of his eldest brother, and had been ever at feud with his father. He collected a body of Mohmands, who were joined by sundry Safis, Bajourdis, and others, and came down to British frontier on 5 December, 1863. Here at the Fort of Shabkadr a Captain J. M. Earle was in command. Firing being heard on the Abanzai road, Captain Earle moved out with fifty-five of the Bengal cavalry and ninety-six native infantry. The enemy, numbering about five hundred, had posted three hundred of their force on a little hill. Earle advanced against them with his infantry in skirmishing order, and his cavalry on his flanks. The latter charged from both flanks and disposed of some six or seven of the enemy killed and about twenty wounded. During the charge Lieutenant Bishop who led on the left, was mortally wounded, and a *sowar* also, though not seriously.

The enemy retired to the crest of one of the hills of the first range before the infantry came up. As this was the wrong side of our frontier Captain Earle did not pursue them but retired and the enemy did not follow him up. Reinforcements were now sent to Shabkadr from Peshawur under Lieut.-Colonel Jackson of the 2nd Bengal cavalry.

On 7 December there was another affair. The enemy had taken possession of the ridge in front of Shabkadr and had to be driven out. This was done, but the engagement was warm while it lasted, especially on the left front of the line, where Lieutenant A. FitzHugh of the 4th Sikhs was closely engaged in keeping the enemy at bay. When evening came on Lieut.-Colonel Jackson withdrew his force. As it fell back it was perpetually fired on at long range. When darkness fell a party of the enemy took possession of a village about eight hundred yards from the fort and had to be shelled out. On this day the casualties were two sepoys of the 4th Sikhs killed, one jemadar and one sepoy of the 4th Sikhs wounded, and two of the 8th Native Infantry. At the end of the month Mohammad Khan took up a position at Regmiana, a small village about five or six miles distant from the fort. His following then did not number more than four hundred men, but he was soon after joined by three hundred

others under Nauroz Khan, the son of Saadhat Khan. These last occupied the village of Chingai to the north-east of Regmiana. Mul-
lahs now appeared on the scene from Afghanistan and stirred up
the tribesmen to wage a religious war. At first their success was not
great, but by the end of the month the rebel force was augmented
to the number of three thousand eight hundred.

On 1 January three bodies arrived in addition, and Mohammad
Khan by the night of 2 January found himself at the head of five
thousand six hundred men.

He now determined to move out and try conclusions with the
British force at Shabkadr. But in the meanwhile the fort had been
considerably reinforced, and was now garrisoned by one thousand
seven hundred and fifty-two men. The force was thus composed: D
Battery, 5th Brigade Royal Horse Artillery, 3 guns, 2 officers, and
49 men; 7th Hussars, 5 officers and 140 men; 3rd Battalion Rifle
Brigade, 27 officers and 691 men; 2nd Bengal Cavalry, 5 officers
and 231 men; 6th Bengal Cavalry, 1 officer and 95 men; 2nd Gurkhas, 7 officers and 453 men; and 4th Sikhs, 1 officer and 93 men.
Total, 48 officers and 1752 men. The whole were commanded by
Colonel Alexander Macdonell, C.B., of the Rifle Brigade.

On the morning of 2 January the Mohmands and other tribesmen having collected at Regmiana, marched out from the gorge on
the north-west of the fort and formed up on the plateau in front of
it, where they had some five thousand matchlock men arrayed, and
forty horse. They were ranged in the form of a crescent.

Colonel Macdonell then occupied the village of Beri Shamberi,
in the front of his centre, with a company of the Rifle Brigade and
one of the 2nd Gurkhas. By this he hoped to entice the enemy
down from the higher ground. He had also stationed a squadron
of the 7th Hussars on the Michni road to draw the skirmishers of
the enemy similarly. This plan was in part a success, for the right
wing of the enemy advanced. The cavalry thereupon also advanced,
turned and gradually folded the enemy's right on the centre. The
three guns of the Horse Artillery which were posted in front of
Beri Shamberi now opened fire and with some success plied the
retiring wing of the enemy.

The 7th Hussars continued their turning movement and charged
the enemy no fewer than three times. Colonel Macdonell then sent

CIRCA 1864

forward the 3rd Battalion of the Rifle Brigade in skirmishing order. The enemy were then driven across the border and the British returned towards evening to Fort Shabkadr. The enemy's losses were believed to be some forty killed and as many wounded. The British loss amounted to two killed and seventeen wounded, four horses killed and fifteen wounded. Of these casualties the 7th Hussars furnished nine, the 2nd Bengal Cavalry had eight *sowars* wounded. Total causalities, nineteen. (See Frontispiece.) Of this charge it has been written:

> This is the only occasion on which British Cavalry have had an opportunity of distinguishing themselves as a body in Indian border warfare, but the one and only charge to their credit in this class of campaign ranks high among the achievements of the mounted arm.

Shabkadr Fort itself was built by the Sikhs. It stands on a mound, and has walls fifty feet high, so is practically impregnable to any force without artillery. The only other opportunity for cavalry in Indian border warfare occurred in 1897, when two squadrons of the 13th Bengal Lancers had the chance of repeating the success of the 7th Hussars in 1863. They had been escorting the guns and were sent round the enemy's flank quite close to the Shabkadr Fort (8 August). The enemy were endeavouring to cut off the 1st Battalion Somersets and Punjabis (20th) from the fort. They (the 13th Bengal Lancers) charged—the enemy broke and were pursued in disorder as far as the ground would permit.

The Indian Medal, with a clasp for the North-West Frontier, was granted to all survivors in 1876.

The Regimental Record gives the name of the officer in command McDonald, but this is an error, it being Macdonell. Colonel Alexander Macdonell, at that time a C.B., rose later to the rank of General (1 April 1882), Colonel Commandant of the Rifle Brigade (24 January 1886), and was created a K.C.B.

On 24 March 1864 the Regiment was inspected by Major-General Garvock, C.B. Two days later the 7th Hussars, under the command of Colonel Scudamore, C.B., marched from Peshawur for Kamilpore, where they arrived on the 31st.

On 23 April a squadron (A and F Troops), under command of Lieutenant the Hon. W. Harbord, marched to Rawal Pindi and only rejoined the headquarters of the Regiment on 29 November.

This took place when the headquarters were marching through Rawal Pindi *en route* for Sealkote. The Regiment left Kamilpore on 25 November, and arrived at Sealkote on 13 December. On 23 December it was inspected by Major-General Cunynghame, C.B.; on 12 April 1865 by Major-General Haly, C.B., and also by the same officer on 24 November.

Major-General Rainier was the inspecting officer for 1866, the inspections taking place on 28 March and 23 November, and also for 18 March and 6 December 1867.

Yet another alteration in the establishment of the Regiment is to be recorded. The instructions were dated 27 May 1867. By this the strength was fixed at eight troops, the only alterations being the addition of a sergeant cook, a schoolmaster, and a bandmaster. The total of privates became four hundred and twenty-six and the number of horses four hundred and forty.

Major-General Rainier, now commanding the Lahore Division, again inspected the Regiment in 1868—on 13 March and 22 October. On 1 April 1868 Colonel A. Scudamore, C.B., retired on half-pay and Lieutenant-Colonel H. A. Trevelyan assumed the command of the Regiment. On 21 April the Commander-in-Chief, Sir William Mansfield, inspected the 7th and expressed the highest approbation of both the appearance and efficiency of the officers and men. On 10 November three hundred and thirty-seven Snider breech-loading carbines were received from the Ferozepore Arsenal for the use of the Regiment, and the whole of the old Sharp's breech-loading carbines given into store.

From 8 March to 13 April 1869 the headquarters of the Regiment, consisting of three field officers, six captains, three lieutenants, five cornets, one adjutant, one quartermaster, two assistant surgeons, one veterinary surgeon, twenty-eight sergeants and two hundred and seventy-four rank and file, marched for Lahore to be present at the reception of Shere Ali Khan, the Ameer of Cabul The Regiment reached its destination on 13 March, and remained encamped on the racecourse until the return of the Ameer from Umballa. The Regiment left Lahore on 9 April.

During 1869 the following inspections took place. On 20 April by Brigadier- General J. E. Goodwyn, C.B., commanding the

SHERE ALI KHAN & SUITE

Sealkote Brigade and on 13 November by Major-General Haly, C.B., commanding the Jhelum Division.

On 19 January 1870, by telegraph, orders were received for the 7th Hussars to proceed to Lahore immediately *en route* for England. The Regiment was opened for volunteering, and fifty-eight non-commissioned officers and men transferred their services to regiments serving in India. The Regiment marched from Sealkote on 23 January and reached Lahore on 27 January. Horses, arms, and saddlery were handed over on the next day, and the Regiment immediately proceeded by rail to Bombay, and after halting at Jubbulpore, Nagpore and Deolali they arrived at Bombay on 27 February. These halts were occasioned by the postponement of the embarkation from 15 February to 28 February. However, on the day of arrival at Bombay the Regiment was at once embarked on the *Euphrates*. Before sailing on the morrow the inevitable inspection took place. Unfortunately another delay occurred in consequence of the *Euphrates* colliding with another vessel in Bombay Harbour. This obliged the troopship to return, and not until 5 March were the repairs which were required completed.

Suez was reached on 18 March. Here on the 20th the troops on board disembarked and proceeded by rail to Alexandria, where the Regiment immediately re-embarked, this time on H.M. troopship *Crocodile*. The *Crocodile* sailed from Alexandria on 31 March and reached Portsmouth on 8 April. The 7th disembarked on the same day, and having been inspected the Major-General Commanding the Southern District, proceeded by rail to their old station York, at which city they duly arrived on 9 April.

CHAPTER 6

Home Service
1871

The Regiment having handed over its horses prior to leaving India, was on arrival at York practically dismounted. Two hundred and seventy horses were, however, speedily transferred to them from the following regiments: Second Dragoons 24, 3rd Dragoon Guards 44, 6th Dragoons 44, 8th Hussars 14, 9th Lancers 14, 10th Hussars 16, 12th Lancers 14, 13th Hussars 13, 14th Hussars 14, 17th Lancers 44, Cavalry Depôt 29.

On 3 May, in accordance with authority from the War Office, the establishment of the Regiment was fixed, as from 1 April, to consist of the following numbers: Seven troops: 1 colonel, 1 lieutenant-colonel, 1 major, 7 captains, 7 lieutenants, 3 cornets (A. C. 1870, Clause 95), 1 paymaster, 1 adjutant, 1 riding-master, 1 quartermaster, 1 veterinary surgeon, 1 regimental sergeant-major, 1 trained bandmaster, 1 regimental quartermaster-sergeant, 7 troop sergeants-major, 1 paymaster-sergeant, 1 armourer-sergeant (unless supplied by the Corps of Armourers), 1 farrier-major, 4 farriers, 1 saddler-sergeant, 1 hospital-sergeant (unless supplied by the Army Hospital Corps), 1 orderly room clerk, 1 sergeant-instructor in fencing, 1 sergeant cook, 21 sergeants, 1 trumpet-major, 7 trumpeters. 21 corporals, 9 shoeing-smiths, 2 saddlers, 1 saddle-tree maker, 374 privates; total 483. Horses 300. Attached, 1 surgeon, 1 assistant surgeon. Comparison with previous establishments will show that there were certain changes, some additions, and one or two grades dropped.

On 7 May the Regiment was inspected by Major-General Sir John Garvock, K.C.B., Commanding the Northern District.

The changes in the establishment for this year were not, however, ended. By an authority from the War Office, dated 24 August, the establishment was now augmented, as from 13 August:

Rank and File		Horses
Corporals	28	
Shoeing-smiths	10	
Saddlers	3	
Saddle-tree maker	1	
Privates	448	
Total	490	350

On 21 September the Regiment was inspected by Major-General H. D. White, C.B., Inspector-General of cavalry in Great Britain.

Meanwhile on 20 July recruiting orders were received. By 17 December hundred and eighty-seven recruits were raised, eighty-one of whom were enlisted at regimental headquarters. This was a very creditable percentage for five months, and bears testimony not only to the activity of the Regiment in recruiting, but also demonstrates its popularity as a corps: a popularity gained by the good conduct of the men, their smartness, and their self-respecting demeanour.

Here two pages of the *Manuscript Record* have been securely pasted together. The next entry is dated February 1871, and records the retirement of Colonel H. A. Trevelyan from the service. He was succeeded in the command of the Regiment by Lieut.-Colonel Robert Hale on 21 February.

During the month of May the Regiment marched from York as follows; their destination being the North Camp, Aldershot:

	Departure	Arrival
E and F Troops	6 May	22 May
A and B Troops	11 May	26 May
C and D Troops	12 May	27 May
G and H Troops and Head-quarters under the command of Lieut.-Colonel R. Hale	16 May	31 May

The dismounted portion of the Regiment left in two divisions on the 6th and 16th Their date of arrival is not given, neither is it stated whether they travelled by train or marched.

The Regiment remained in camp during the summer.

On 2 June they were inspected by Lieut.-General Sir J. Hope Grant, K.C.B., who then commanded the Aldershot Division. Of the officers who had been under the command of Sir Hope Grant during the Mutiny, one alone remained in the Regiment, in the person of Lieut.-Colonel Robert Hale.

The inspection by the Inspector-General of Cavalry, Commanding the Cavalry Brigade at Aldershot, Major-General Sir T. McMahon, C.B., took place on 21 June.

During September the Regiment took part in the Autumn Manoeuvres of 1871 under the command of Lieut.-Colonel R. Hale. The 7th Hussars formed part of the Third Division, which was commanded by Major-General Sir Charles Stavely. The Regiment was formed in three squadrons—its strength 19 officers, 25 sergeants, 7 trumpeters, and 259 rank and file; 30 officers' chargers and 234 troop-horses. A review of the whole of the troops engaged in the Manoeuvres was held by H.R.H the Duke of Cambridge, Commander-in-Chief, on 22 September. The Right Hon. Edward Cardwell, Secretary of State for War, was also present on the occasion.

The 7th Hussars, who had been under canvas since May, during the month of October moved from the North Camp into the West Cavalry Barracks.

On 16 October 1871 the annual inspection was held by Major-General Sir Thomas McMahon, C.B.

On 1 November the rank of Cornet was abolished; that is to say, those officers hitherto styled Cornets were henceforward to be designated Sub-Lieutenants.

The events of 1872 were as follows: in April, pantaloons and boots were issued to the Regiment instead of the leathered overalls hitherto worn. From the point of view of appearance this change was certainly an improvement.

On 3 April Sir Thomas McMahon, C.B., made his half-yearly inspection. In the *Regimental Record* an establishment of the Regiment is given dated 1 May, 1872. From this we gather that there were now 8 captains instead of 7, 12 lieutenants and sub-lieutenants instead of 7 lieutenants and 3 cornets; 1 more of each of the following: troop sergeant-major, farrier, trumpeter, shoeing-smith, and saddler; 3 more sergeants and 11 more corporals. Thus there were 28 officers, 59 non-commissioned officers and trumpeters, and 447 rank and file; total, 534 all ranks; and 320 troop-horses.

On 5 July Her Majesty Queen Victoria visited Aldershot, on which occasion the whole division was reviewed and marched past.

His Royal Highness the Prince of Wales (afterwards King Edward VII.) was also present, as well as H.R.H the Duke of Cambridge, Commanding-in-Chief.

The season for the Autumn Manoeuvres for 1872 arrived in August, and they duly took place on Salisbury Plain. For this purpose the A and B Troops were broken up and the Regiment formed in three squadrons. The strength was 22 officers, 30 sergeants and 288 rank and file; 40 officers' chargers and 260 troop-horses. The remainder of the Regiment, with the women and children, were formed into a depôt.

The Regiment left Aldershot on 14 August, one squadron marching *via* Alton and two squadrons and headquarters *via* Basingstoke. The 7th were billeted at night until they arrived at Blandford on the 17th, where the camp was. The Light Cavalry Brigade, of which they formed a part, consisted of the 7th and 10th Hussars, the 12th Lancers and D Battery, B Brigade Royal Horse Artillery, Colonel Valentine Baker (10th Hussars) being in command of the Brigade, which part of the Southern Army Corps. The Southern Army Corps was commanded by Lieut.-General Sir John Michell, K.C.B. The regimental transport, which consisted of six wagons containing camp equipment, one supply wagon, one forge cart and one water cart, all horsed by the Regiment (thirty-two horses in all), accompanied the 7th throughout the manoeuvres.

On 12 September, the manoeuvres having been concluded, the whole of the troops, i.e. both the Northern and the Southern Army Corps, numbering 30,492 of all ranks, 84 guns and 5749 horses, were inspected on Beacon Hill and marched past H.R.H the Duke of Cambridge, Commander-in-Chief. His Royal Highness the Prince of Wales (afterwards King Edward VII.) and the Right Hon. Edward Cardwell, Secretary of State for War, both being present on the occasion.

On the following day the Southern Army Corps was broken up, and on 14 September the 7th Hussars left camp near Amesbury and marched for Hounslow. Here they went into quarters on 18 September, sending out detachments to Hampton Court and Kensington, under the command of Captains Peel and Hunt respectively. The depôt, with the women and children, had previously arrived at Hounslow on the 16th.

On 26 and 27 September the Regiment was inspected by Major-General Sir Thomas McMahon, Bart., C.B., Inspector-General of Cavalry.

For the month of October there is no event to record.

On Lord Mayor's Day, 9 November 1872, a squadron (fifty men with the trumpeters and kettledrums), under the command of Captain Gist, proceeded to London to act as escort to the Lord Mayor from the Guildhall to Westminster and back, returning to Hounslow in the evening.

Seventy recruits and fifty-six horses joined the 7th Hussars during 1872, fifteen of these horses being received from the 10th Hussars, who were ordered for service in India.

In April 1873 the Regiment marched from Hounslow to Wimbledon for the purpose of taking part in the Easter Monday Volunteer Review, which was held there under the command of His Serene Highness Prince Edward of Saxe-Weimar, C.B., Commanding the Home District. It may be noted that the Regiment had never previously attended a Lord Mayor's Show or taken part in a Volunteer Review.

During June the Shah of Persia paid a visit to this country, and a travelling escort of the 7th, under the command of Lieutenant Harold Paget, was quartered at Regent's Park on 20 June. These, with the detachment stationed at Kensington, performed the escort duties during the stay of His Majesty.

The whole Regiment was present at the review held at Windsor by Her Majesty Queen Victoria in honour of the Shah on 24 June. At the conclusion of the ceremony the Regiment returned to Hounslow.

On 6 August, and again on 22 and 23 August, the 7th Hussars were inspected. On the first occasion Major-General Prince Edward of Saxe-Weimar, C.B. Was the inspecting officer, and on the two last Major-General Sir Thomas McMahon. Bart. C.B. Both officers expressed unqualified approbation of the high state of efficiency and smartness in which they found the Regiment.

Being ordered to change quarters from Hounslow to Maidstone, the Regiment marched on 1 and 2 September 1873, and arrived at their new station on the 3rd and 4th; a detachment of three troops (A, B, and C), under command of Major H. D. Burnell, being at the same time ordered to Shorncliffe.

1 October 1873.—By Royal Warrant of this date 'the ration of bread and meat was granted free to non-commissioned officers and men with an increase of a halfpenny *per diem* to their pay.'

16 December 1873.—The Regiment was inspected by the Colonel (formerly Lieut.-Colonel) General Charles Hagart, C.B., who expressed his approval of the very satisfactory state in which he found his old Regiment.

During the year one hundred and six recruits and twenty-two remount horses joined the Regiment.

At 6 p.m. on 25 January 1874, owing to an outbreak of fire, the detachment quartered at Shorncliffe suffered the loss of fourteen troop-horses, with the greater part of their saddlery. The fire took place in two of the wooden stables, both of which were consumed. It appears that the outbreak was discovered by the fireman on the look-out at about 6 p.m. one Sunday evening. Luckily there was no wind, as otherwise, or had there been even a breeze from the south-west, very much more damage must inevitably have taken place. The fire engine and brigade were soon on the spot and set to work, but a lack of water hampered their efforts. The scene of the fire was about the centre of the line of stables. Twenty-four horses were inside, but despite all endeavours, which were most gallantly made, to persuade them to come out only ten could be extricated and the remainder, together with the saddlery, were burnt. The iron screens between the huts protected the adjacent stables. All the other horses in the range were turned out, but becoming terrified, as horses will in the presence of fire, stampeded. Some galloped through Sandgate as far as Folkestone, and others in the opposite direction. It is needless to add that every effort was made to save the lives of the unfortunate animals both by the fire brigade and also by the men of the Regiment and those of the other corps in the camp. The stables were built of wood, with slate roofs. It will be remembered that a far more disastrous fire had taken place at Aldershot a few years previously, in the course of which several horses that had been successfully brought out blindfolded, when the wrappings were removed wheeled round and, trotting back into their old stalls, perished in the flames.

At Aldershot the stables in question were of a temporary character, being roofed with canvas, with very inflammable walls.

During the month of March a squadron consisting of forty files, under command of Major the Hon. Walter Harbord, proceeded from Maidstone to Gravesend to furnish an escort for Their Royal Highnesses the Duke and Duchess of Edinburgh on their first visit to England after their marriage.

14 April, 1874.—On this date the 7th (Queen's Own) Hussars had the distinguished honour conferred on it of having His Royal Highness Prince Arthur, K.G., K. T., &c, &c. (now Field-Marshal H.R.H the Duke of Connaught), gazetted as one of its captains.

14 May 1874.—D and G Troops, under the command of Captains A. Peel and F. Shuttleworth respectively, marched from Maidstone to Aldershot for the purpose of escorting His Imperial Majesty the Emperor of Russia on the occasion of his visit to the camp. A strong squadron, under the command of Lieut.-Colonel R. Hale, also proceeded to Gravesend, which was intended to act as escort to the Emperor on his landing, but owing to the Imperial yacht grounding at Flushing the landing in England did not take place until later in the day, and not at Gravesend but at Dover. The squadron was in consequence ordered to return to Maidstone.

On 17 May 1874 the headquarters and three troops of the Regiment, which included those at Shorncliffe, marched for Woolwich and took part in a review held in honour of the Emperor of Russia by H.R.H. the Field-Marshal Commanding-in-Chief. Two days later the Regiment furnished a travelling escort for the Emperor from Buckingham Palace to New Cross. B Troop, under the command of Captain H.R.H. Prince Arthur, formed the escort for the Emperor to and from the Arsenal at Woolwich to the Common on the occasion of the review.

On the conclusion of the visit of His Imperial Majesty the Regiment returned to Maidstone, *via* Gravesend, where it escorted the Emperor to the landing stage on his departure.

The Shorncliffe detachment was now withdrawn and three troops were quartered at Woolwich.

Here on 24 June the Regiment was inspected by Major-General Sir Alfred Horsford, C.B. (the Brigadier Horsford of the Mutiny campaign). General Horsford expressed himself as 'highly pleased at its condition.'

30 June, 1874.—The Regiment, under the command of Lieut.-Colonel R. Hale, marched to Aldershot and took part in the summer drills.

They remained there in camp under canvas, the horses in temporary stables partly in the North Camp and partly in the neighbourhood during the drills, after which they proceeded to Norwich by route march, detaching a squadron (E and F Troops) at Ipswich.

During August the Regiment was inspected by Major-General Sir Edward Greathead, commanding the Eastern District at Norwich.

On 10 and 11 September Major-General Sir Thomas McMahon, Bart., C.B. Inspector-General of Cavalry, held his annual inspection of the Regiment.

During 1874 seventy-five recruits and forty-four remount horses, among the latter being fourteen transferred from the 9th Lancers, who were about to proceed to India, joined the Regiment.

In May 1875 H Troop, under command of Captain H. A. Reid, marched from Norwich to Liverpool, and the squadron at Ipswich proceeded to Manchester to be quartered there. These detachments furnished escorts for the Sultan of Zanzibar on the occasion of his visit to Liverpool and Manchester.

During July the D Troop was sent to Liverpool and headquarters and the remaining troops left Norwich for Manchester, where they arrived on 6 August.

On 13 August a squadron (strength seventy files) with the band, under the command of Captain F. Massy Drew, marched to Sheffield on escort duty during the visit of Their Royal Highnesses the Prince and Princess of Wales. The squadron returned to Manchester on the 10th.

His Royal Highness the Duke of Connaught, K.G., K.T., &c., &c, was on 7 August promoted to the rank of Major in the Regiment.

On 24 and 25 August the annual inspection of the Regiment by Major-General Sir Thomas McMahon, Bart., C.B., Inspector-General of Cavalry took place.

During September new regulations were received relative to the granting of good conduct medals, and thirty-eight non-commissioned officers and men were recommended for the same. These medals were subsequently issued. One hundred and seven recruits and thirty-six remount horses joined the Regiment in 1875.

During May a Royal Warrant, with effect from 1 April 1876, was promulgated granting an increase in the rate of pay of all ranks of non-commissioned officers ranging from *2d.* to *6d. per diem.* Pay was also granted under this warrant to lance-sergeants and lance-corporals, and the system of deferred pay was introduced, viz. *2d.* a day to all ranks, not payable until after the discharge of the soldier. An increase in the rate of lodging-money was also granted to staff-sergeants.

Major His Royal Highness the Duke of Connaught, who had been absent from the Regiment while holding the appointment of Assistant Adjutant-General at Gibraltar since October 1875, was now appointed a personal *aide-de-camp* to Her Majesty, and rejoined the 7th Hussars, assuming the command of the squadron at Liverpool.

The Regiment was ordered to Edinburgh and Hamilton in June 1876. The squadron for Hamilton marched by separate troops, C Troop with Captain Thomson on 30 June, B Troop with Captain Drew on 3 July.

The squadron for Liverpool marched thence on 10 July under H.R.H. the Duke of Connaught and moved on 29 July. The remainder of the Regiment moved by separate troops from Manchester during the month, the last party including headquarters, under the command of Colonel R. Hale, reaching Piershill Barracks on 9 August.

During the visit of Her Majesty Queen Victoria from 16 to 18 August the Regiment performed the following duties. Travelling escorts on arrival and departure; royal escort at the unveiling of the statue of H.R.H. the Prince Consort in Charlotte Square, the remainder of the Regiment lining the streets, for which duties the squadron from Hamilton was also brought to Edinburgh, returning after the conclusion of the proceedings. At the ceremony of unveiling the statue Major H.R.H. the Duke of Connaught commanded the escort.

During this month the issue of one cart per squadron as an article of equipment was introduced for use on the line of march and for ordinary regimental transport in quarters.

The annual inspection of the 7th Hussars by Major-General R. Wardlow, C.B., Inspector-General of Cavalry, took place on 24 and 25 August.

A considerable alteration in the course of musketry instruction was made during this year. The mounted practice was abolished, and dismounted practice with horses was substituted in lieu thereof. There were also other changes as to distance of ranges.

During September, it being found that the establishment of the Regiment was below strength, recruiting was opened in the London district. Fifty recruits were obtained in the course of the months of September and October.

Major-General Ramsey Stuart, C.B., commanding the forces in Scotland, inspected the Regiment and was pleased to express his great satisfaction at the appearance of the men and horses and with the general result of his inspection.

During October, on the occasion of His Royal Highness the Prince of Wales laying the foundation stone of the new post office at Glasgow, the Regiment proceeded to that city, returning to Edinburgh and Hamilton on the conclusion of the ceremony.

A Royal Warrant was issued limiting the number of lance-sergeants and lance-corporals who were to receive extra pay for the acting rank. There were in future to be four of the former and twelve of the latter to each cavalry regiment.

The last event of the year was the issue of fresh regulations as to the numbers to be borne on the strength of the married establishment, the effect being a reduction from seven to four per cent. Of the rank and file.

Ninety-six recruits and twenty-six remount horses joined the Regiment during the year 1876.

In May 1877, during the sitting of the General Assembly of the Church of Scotland at Edinburgh, the 7th Hussars furnished escorts, &c. daily to His Grace Her Majesty's High Commissioner.

During June the establishment of the Regiment, which had remained unchanged since 1 May 1872, was slightly altered.

There were now twenty-seven officers instead of twenty-eight, the paymaster no longer appearing in the list. The rest of the Regiment seems as before, except that the number of privates is reduced from four hundred and three to four hundred. This gives the following totals: four hundred and forty-four rank and file, five hundred and thirty all ranks, and three hundred and seventeen troop-horses.

In June 1877 the Regiment was ordered to Ireland, its quarters being fixed at the Island Bridge Barracks, Dublin.

The troops marched in three equal divisions from Edinburgh as follows: three troops, under the command of Lieut.-Colonel Burnell, left on 27 June by route march, embarked at Greenock on 30 June, and disembarked at Dublin on 2 July.

Two troops and headquarters, under command of Colonel Hale, marched from Edinburgh on 2 July, embarked at Greenock on 5 July, and disembarked on 7 July.

One troop marched from Edinburgh on 5 July, and the two troops from Hamilton under the command of Captain Drew marched on 7 July. The latter party embarked at Greenock on 10 July and disembarked on 11 July. The Regiment was conveyed to Ireland in H.M.S. *Assistance,* and the journey was performed without any casualties to men or horses. 14 August 1877.—A troop, under command of Captain Hunt, was despatched by special train to Lurgan in aid of the civil power. The troop remained at that place for two days, when they returned by route march to Dublin, rejoining headquarters on 24 August. Ireland was now much disturbed owing to the action of the Land League. The troops, both cavalry, artillery, and infantry, were busily engaged in consequence. There was much marching and counter-marching about the country. Later, considerable reinforcements both of men and material arrived from England. This state of things continued off and on during the remainder of the stay of the Regiment in Ireland.

The 7th Hussars being at this time under its fixed establishment, recruiting was opened in the London district. Seventy men were thus obtained during the months of July and August.

25 and 26 September 1877.—The Regiment was inspected by Major-General Seymour, C.B., Inspector-General of Cavalry in Ireland, who was pleased to express his satisfaction at the result.

In the month of December authority was issued for the transfer of cavalry soldiers in excess of the establishment who had served four years in the ranks, to the Army Reserve for the purpose of completing their service therewith. subject to the same rules laid down for the transfer of the men of infantry regiments.

We here insert a reproduction of the only portrait of H.R.H. the Duke of Connaught in the uniform of the Regiment which

Major H. R. H. the Duke of Connaught, K. G., K. T., K. P., G. C. M. G.

we have been able to obtain, and it has also the advantage of his signature and date.

In January 1878 an order was issued by H.R.H. the Field-Marshal Commanding-in-Chief directing that all discharges from the army 'by indulgence' should be suspended until further notice. During April D Troop, under command of Captain A. Peel, marched from Dublin to the Curragh Camp, where they took up the vedette duties at that station (3 April).

Her Majesty Queen by Royal Proclamation having declared that a case of 'great emergency' existed, the transfer of men to the Army Reserve was suspended until further notice.

On 1 April there was a slight change in the establishment, four second-lieutenants being added. The strength of the Regiment now stood at thirty-one officers, fifty-two non-commissioned officers, a proportionate number of other ranks, and five hundred and seventeen privates; the total becoming five hundred and sixty-one rank and file—six hundred and fifty-two of all ranks, with four hundred and eighty troop-horses.

This was an increase of four officers, one hundred and eighteen men and one hundred and sixty-three troop-horses.

By route march on 28 May the Regiment, under command of Colonel R. Hale, proceeded from Island Bridge Barracks, Dublin, to Newbridge Barracks, Co. Kildare; the dismounted party, the women and the children proceeding thither by rail on the same day.

The Snider carbines were now withdrawn and returned into store, Martini-Henry carbines being issued in lieu thereof.

During June, in consequence of the increase in the establishment already noted, recruiting was opened in all the sub-districts, and seventy-eight recruits joined during the month Fifty-eight others were obtained during July, after which recruiting ceased.

The Regiment was now forty men in excess of its establishment.

It was now decided by the War Office that remount horses were not to be purchased at present to complete the establishment. The Regiment was inspected by Major-General Seymour, C.B., Inspector-General of Cavalry in Ireland, on 15 and 16 July.

Transfers to the Army Reserve were now authorised to be carried on as usual, the order for suspension of the same issued in April having been cancelled. The order suspending discharges by purchase and other indulgence was also cancelled. It was likewise notified that men who had completed eighteen years' service and were desirous of taking their discharge would be recommended for the 'indulgence to modified pension' under Article 1279, Royal Warrant, 1 May 1878.

The establishment of the Regiment during August was again changed and it reverted to the strength given under date June 1877.

4 September 1877.—Major-General Glyn, C.B., Commanding the Dublin District, inspected the Regiment.

During the year one hundred and sixty-nine recruits and twenty-six young horses joined the Regiment.

Colonel R. Hale, who had completed a continuous service of twenty-seven years with the 7th (Queen's Own) Hussars, during which he had been its Lieutenant-Colonel for more than eight years, now retired upon half-pay. He took leave of the Regiment at a dismounted parade in review order on 22 June. On retirement he was appointed an *aide-de-camp* to the Field-Marshal Commanding-in-Chief. The loss of Colonel Hale was much regretted by the Regiment with whom he had served so long. He was succeeded in the command by Major and Brevet-Colonel Hugh D'Arcy Pegge Burnell.

On 30 July the death took place of General C. Hagart, C.B., the Colonel of the Regiment. General Hagart died at his residence at Guildford. His successor was General Henry Roxby Benson, C.B.

It was gratifying to the Regiment to receive from the Commander of the Forces in Ireland a special notice 'for the very excellent sketches and reports' sent in during the season of 1879.

The annual inspection of the 7th Hussars by Major-General Seymour, C.B., Inspector-General of Cavalry in Ireland, took place on 5 and 17 August. General Seymour expressed himself as 'perfectly satisfied, and much pleased' with everything he had seen.

18 October 1879.—The Regiment was called upon to furnish ten volunteers for the 10th Hussars.

On 5 November the Regiment marched from Newbridge to Cahir, furnishing the following detachments: one and a half troops to Gort, under Captain Phipps; two troops to Fethard under Captain Reid, one troop to Waterford under Captain Hunt. Later C Troop, under a subaltern, was sent to Clogheen.

As from 1 October the establishment of the Regiment was by Special Army Order dated 22 November fixed as follows: Twenty-seven officers, fifty-one non-commissioned officers, four hundred and ninety-eight privates; total rank and file, five hundred and forty-two. Total all ranks, six hundred and twenty-eight. Total troop-horses, three hundred and seventy-nine. In accordance with the above order, recruiting was opened in the various districts of Great Britain and Ireland, and during the year 1879 seventy recruits and forty-eight young horses joined the 7th Hussars.

During March 1880 the Regiment was called upon to furnish fifteen volunteers for the 15th Hussars.

In April 1880 a fresh pattern forage-cap with a figured crown was approved for all Hussar regiments.

4 May 1880, the detachment at Fethard, under command of Captain Shuttleworth, marched to Cork.

19 May, the Gort detachment, under Captain Phipps, marched to Limerick.

28 June, the Clogheen detachment, under Lieutenant Mynors, marched to Cork and joined the detachment there.

During the month of July several changes took place: the three troops at Cork marched, one under Captain Shuttleworth to Fethard, one under Lieutenant Viscount Lumley to Limerick, and one under Lieutenant Mynors to Cahir. The Fethard detachment was also strengthened by half a troop from Limerick, and the detachment under Captain Paget was again furnished to Clogheen from headquarters.

The annual inspection by Lieut.-general H. Hamilton, C.B., commanding the Cork District, was held on 20 August, and that by Major-general C. Fraser, V.C., C.B. Inspector-General of Cavalry in Ireland, took place on 11 and 12 October. General Fraser, it will be remembered, had served in the Regiment from 1847 to 1859, and won his V.C. at the passage of the Raptee. He expressed himself as 'much pleased' with everything he saw at the inspection.

During the month orders were received to purchase thirty-five young horses in excess of the establishment to replace thirty-five trained horses which it was proposed to transfer to the 15th Hussars. Fifteen volunteers were now called for to be sent to the same regiment during November. A good many movements took place in the course of the month of December 1880 in consequence of the seriously disturbed state of the country. A detachment was ordered to be sent to New Ross, and accordingly forty non-commissioned officers and men were selected from Fethard, Clogheen, and headquarters, and were posted at that station under the command of Lieutenant Ridley. All leave for officers and non-commissioned officers was suspended, and those absent were recalled.

The detachment at Limerick was strengthened by forty men under Captain the Hon. A. Byng, who proceeded thither by route march.

On 4 December a party of forty rank and file marched to New Pallas from Limerick under command of Captain Phipps; to render assistance to the Constabulary authorities, but as no disturbance took place the party returned to Limerick on the same day.

On 23 December it was ordered that where cavalry in Ireland were stationed without infantry the carbine service ammunition issued should be increased to fifty rounds per man. At this time munitions of war were being sent into the country in large quantities by the Government, among which buckshot largely figured. It was this fact which furnished a nickname for a certain statesman responsible therefore, he being known from one end of the island to the other as Buckshot Forster.

Only a few days previously an American vessel laden with arms for the Irish malcontents had put in at Los Passages, where it was joined by forty *fenians*, who sailed for Ireland and succeeded in landing.

On 24 December a party of twenty non-commissioned officers and men, under the command of Captain the Hon. A. Byng, marched from Limerick to Kilmallock station in order to escort thence to Kilfinane some regimental stores and baggage, &c, belonging to a detachment of the 48th Regiment. Finding the accommodation insufficient and the behaviour of the inhabitants too threatening. Captain Byng, in order to avoid running any risk of a conflict, did not put up in billets as had been at first intended, but marched back to Limerick on the same night, arriving there at 3.15 a.m. On Christmas morning after having been in the saddle uninterruptedly for twenty-two hours and having marched sixty-six miles. This was alluded to at the Land League meeting in Dublin on 28 December, when the 'Carriers and Labourers of Limerick' were raised for refusing to convey military baggage from Kilmallock to Kilfinane.

Several other small detachments were called out at various times during the year to aid the civil power, but in no instance did any actual disturbance take place, though more than once the state of affairs was certainly threatening and the slightest want of judgement on the part of the officers in command or their men might have precipitated a serious conflict.

It undoubtedly speaks well for the tact evinced by the officers and non-commissioned officers and the high state of discipline and good humour of the men of the 7th (Queen's Own) Hussars.

During 1880, one hundred and forty-five recruits and one hundred and seventeen young horses joined the Regiment.

In January 1881 the Regiment was thus distributed in the Cork district: Cahir, two and a half troops and headquarters under command of Colonel Burnell; Limerick, two and a half troops under Captain and Brevet Lieut-Colonel Hugh McCalmont; Waterford, one troop under Captain Hunt; Clogheen a half-troop under Captain Harold Paget; New Ross, a half-troop under Lieutenant Ridley; and Fethard, one troop under Lieutenant Viscount Somerton.

Being ordered to proceed to Natal, South Africa, in February 1881, the Regiment proceeded by route march to Dublin and concentrated there on 21 February, and subsequently embarked.

The record of Natal, Home Service, and Egypt (Camel Corps) will be contained in the next chapter.

CHAPTER 7

Natal & Sudan

1881

At the end of the last chapter we left the Regiment concentrated at Dublin and under orders to proceed to Natal, the date being 21 February 1881. Prior to embarkation the 7th Hussars were inspected by Major-General C. C. Fraser, V.C., C.B., Inspector-General of Cavalry in Ireland, who also addressed them on the occasion of their departure on what pointed at the time to active foreign service. His Excellency the Lord-Lieutenant of Ireland similarly inspected them and addressed them.

The Regiment embarked thus: Headquarters with A, C, D, E, and H Troops, consisting of fourteen officers, one hundred and eighty non-commissioned officers and men, and two hundred horses, sailed in the s.s. *Calabria* from Kingstown on 26 February. Three detachments, consisting of three officers and one hundred non-commissioned officers and men, proceeded to Hounslow, Colchester, and Aldershot respectively, to take over remounts from the 3rd, 4th, and 11th Hussars, and embarked at London on s.s. *Nemesis* and s.s. *France*. The *Nemesis* proceeded to Kingstown to embark a detachment of the Regiment which was awaiting her there. The latter vessel sailed for South Africa on 25 February. The remainder of the Regiment, under the command of Major Francis Massy Drew, embarked on the *Nemesis* on 27 February at Kingstown.

The *Calabria* arrived at Port Natal on 4 April, having lost eight horses on the voyage. The *France* arrived there on the same day. The engines of the *Nemesis* broke down after she had been at sea for a few days, and both men and horses were exposed to extreme heat,

113

in consequence of which thirty-nine horses died. The *Nemesis* at length reached Cape Town on 22 April. Here the troops and horses were transshipped to the *Calabria,* which vessel had returned from Natal after landing the headquarters. The *Calabria* then sailed for Port Natal, where she arrived 5 April. On arrival the Regiment encamped on the Eastern Vlei, Durban, and when the horses had recovered from the effects of the voyage proceeded to Pinetown. again going into camp.

The political position may here be briefly explained. The Boers of the Transvaal during the interval between the British annexation in 1877 and November 1880 had been agitating for independence. When troubles arose between the British and the Zulus, with the exception of Piet Uys and a small band who followed him they had held completely aloof from offering any assistance during the conflict which ensued. The British, under Colonel Evelyn Wood, V.C. (now Field-Marshal Sir Evelyn Wood), at Hlobane were co-operated with by Piet Uys and his followers.

Isandlwana had been fought on 22 January 1879. Rorke's Drift was defended so gallantly on the same date. The enemy were de-feated at Inkanyana on 24 January. Meanwhile reinforcements from England were being hastily sent. Eight hundred men arrived at Pietermaritzburg on 11 March. Next day a British convoy was well-nigh annihilated near Itombi river. Then on 28 March came the affair on the Zlobani mountains and the victory at Kambula. On the 29th the British advance to relieve Eshowe, where a slender force was besieged, began on the same day. On 2 April the Zulus were defeated at Ginghilovo. Colonel Pearson, who was in com-mand at Eshowe, marched out thence on 2 and 3 April.

During May Sir Garnet (afterwards Viscount) Wolseley sailed to the Cape, where he arrived on 23 June, his office being Com-mander-in-Chief in Natal and Governor. The battle of Ulundi, where Cetewayo the Zulu king was totally defeated, was fought on 4 July. Secocoeni was now in revolt. After a harassing campaign his stronghold was captured on 28 November 1879, and that potentate surrendered on 2 December.

During the same month and again in January 1880 the Bo-ers met and claimed independence. They seized Heidelburg on 16 December, and established the South African Republic with

Kruger as President on the next day. On 20 December a party of two hundred and fifty men of the 94th Regiment who were being sent up country were stopped by the Boers at Bronker's Spruit. A conflict ensued and some were killed and wounded; the rest were disarmed and dismissed. Potchefstroom was seized by the Boers 27 December, who retired when artillery was used against them; Colonel Bellairs, who subsequently garrisoned it, being besieged there. On 29 December Captain J. M. Elliot, a prisoner and on parole, was murdered while fording the Vaal. On 30 December the South African Republic was proclaimed by Kruger Joubert, and Pretorius. Troops were now being despatched from England. Then followed the attack on Laing's Nek under Sir George Pomeroy Colley, which was repulsed with heavy loss on 28 January 1881. Similarly on the Ingogo river on 8 February the British met with defeat. Sir Evelyn Wood arrived with reinforcements on 17 February. On 26 and 27 February the disaster at Majuba Hill took place, and there General Colley lost his life.

Sir Frederick Roberts (now Field-Marshal Earl Roberts, V.C., K.G.) was sent out to South Africa. He did not, however, arrive until after the conclusion of the discreditable agreement made with the Boers by the British Government. For forty-eight hours only General Roberts remained in South Africa and then immediately returned home.

This brief table of events will be sufficient to demonstrate why the 7th (Queen's Own) Hussars were deprived of any opportunity of seeing active service in South Africa on this occasion.

Their stay was not protracted; and beyond the fact that they were inspected on two occasions, once on 22 November 1881 by Brigadier-General Drury Lowe, Inspector-General of Cavalry, and the second time on 7 January 1882 by Lieut-General the Hon. L. Smyth, Commanding-in-Chief in South Africa, there is nothing to record. In the month of March orders were received for the Regiment to return home. The horses, with the exception of one hundred and fifty which were transferred to the 6th Inniskilling Dragoons and the Royal Artillery; were sold off by auction at the headquarters, Durban.

Mr. Herbert Compton in *A King's Hussar*,* a volume which gives the reminiscences of Troop-sergeant-major Edwin Mole of

---

* Available from Leonaur under the title *The King's Hussar.*

the 14th Hussars, tells us how the 14th Hussars, while returning from Harrismith after the Majuba campaign, met the 7th Hussars at Pinetown, where they were then in camp. He states that the 14th then lost a large number of men who were drafted into other regiments. Volunteers, partly from various regiments and also from the 7th Hussars, however, filled some of the vacancies. On arrival at Bombay the 14th heard with great regret how both the 7th Hussars at Pinetown and also the 41st Regiment at Harrismith had suffered from an outbreak of enteric.

At page 342 in Mr. Herbert Compton's already quoted book we read that the depôt officers of the 14th Hussars at Colchester, after working hard to render efficient a number of young recruits in order to have a fine body to show the Regiment on its arrival from India, were much disappointed under the following circumstances. On a Sunday after church parade every man in the depôt was ordered to parade in front of the orderly-room. All the men under three months' service belonging to the 14th Hussars were then ordered to be transferred to the 7th Hussars, and all over three months' service who chose to volunteer for the 7th were permitted, nay encouraged, to do so. The 7th were going to India to relieve the 14th. Before evening more than half the depôt had volunteered.

A & E Troops, consisting of seven officers and one hundred and eighteen non-commissioned officers and men, under the command of Colonel Burnell, then embarked on board the R.M.S. *Kinfauns Castle*, and arrived at Portsmouth on 25 April, proceeding thence to Colchester, where they were stationed until further orders. On 31 May the remainder of the Regiment, consisting of B, C, F, G, and H Troops (eight officers and too hundred and twenty-eight non-commissioned officers and men), under the command of Lieut.-Colonel Hugh McCalmont, embarked at Natal on board the R.M.S. *Anglian*, and arrived at Southampton 5 July, proceeding thence to Colchester. From 1 April 1882 the establishment of the Regiment had been fixed by Special Army Circular (dated 1 May 1882) as follows:— 2 lieutenant-colonels, 3 majors, 5 captains, 1 lieutenants, 1 adjutant, 1 riding-master, 1 quartermaster (total 24); 1 regimental sergeant-major, 1 bandmaster (warrant officers 2); 1 quartermaster-sergeant, I sergeant-trumpeter, 1 sergeant-instructor of fencing, 1 paymaster-sergeant, 1 armourer-sergeant, 1 saddler-sergeant, 1 farrier quartermaster-sergeant, 1 orderly room sergeant, 8 troop sergeants-major, 1 sergeant-cook, 24 sergeants, 8 sergeant-farriers (total 49 sergeants); 8 trumpeters, 32 corporals, 8 shoeing-smiths, 3 saddlers, 1 saddletree maker, 366 privates; total rank and file, 410. Total of all ranks, 493; troop horses, 300. Comparison will show that this was a considerable change in the establishment of the Regiment.

The following is the copy of a letter received from the Horse Guards, War Office, dated 20 September, 1882:

Sir, By desire of H.R.H. the Field-Marshal Commanding-in-Chief, I have the honour to acquaint you that Her Majesty has been graciously pleased to approve of the 7th Hussars being permitted to wear on its appointments the word *Dettingen,* in commemoration of the battle fought at that place on the 27th June 1743. I have the honour to be, Sir, your obedient Servant,

R. M. Taylor, A. G.

To Officer Commanding
7th Hussars, Colchester

Orders were now received for a troop to proceed to Cyprus to form part of a remount depot at that station. Accordingly on 2 September 1882 A Troop, consisting of seventy-one non-commissioned officers and men, under the command of Captain T. H. Phipps, embarked on board H.M.S. *Tyne* at Portsmouth. The troop arrived at Alexandria on 15 September, and proceeding to Cyprus disembarked at Limasol on 17 September. The stay of A Troop at Limasol was of a very brief duration, for on 23 October it re-embarked on board the steamship *Palmyra,* and arrived at Portsmouth on 9 November. Here the troop disembarked and proceeded to Colchester to rejoin the Regiment.

On 5 and 6 October the Regiment had been inspected by Major-General Sir F. W. J. FitzWygram.

The following is a copy of a letter received from the Horse Guards dated 30 November 1882:

Sir, A report having been received from the General Officer commanding the troops at Cyprus of the general exemplary conduct of the cavalry and remount depôts during the period of their stay in Cyprus, I am directed by the Field Marshal Commanding-in-Chief to acquaint you that His Royal Highness has expressed his entire satisfaction with this report, and has been pleased to observe that nothing could have been better than the conduct of the men of the several depôts and the attention of the officers; and I am accordingly to request that you will be so good as to communicate the same to the officer commanding 7th Hussars, of whose troop it is stated, their conduct was excellent. I have the honour to be, Sir, your obedient servant, *Wolseley*, A. G.

The General Officer Commanding
Colchester

In the month of September 1884 it was decided to form a Camel Corps for temporary service in Egypt, and a detachment of the 7th Hussars was formed and prepared for embarkation. Two officers, Lieut.-Colonel Hugh McCalmont and Lieutenant the Hon. R. T. Lawley, were selected to accompany the men. The non-commissioned officers and men amounted in numbers to forty-four, viz. three sergeants, one trumpeter, two corporals and thirty-eight privates. Captain Harold Paget was selected as adjutant and quartermaster of the Hussar division, Camel Corps, which embarked at Portsmouth on board the troopship *Australia* on 26 September, and landed at Alexandria on 6 October, proceeding immediately up the Nile to Assouan. Here they remained for a fortnight, after which they marched on camels to Korti.

The Light Camel Corps, as this force was designated, arrived at Korti on Christmas Eve 1884. On 26 October a general order was issued forming the Camel Corps into four divisions to be called respectively the Heavy Camel Regiment, the Light Camel Regiment, the Guards Camel Regiment and the Mounted Infantry Camel Regiment. The total strength of the Light Camel Regiment is given as four hundred and five. Camel marches were ordered not to exceed the rate of one hundred and twenty miles in seven days.

On 13 December Lord Wolseley, accompanied by his personal staff and Sir C. Wilson, who had been at Dongola, left for Korti, where he arrived on 24 December. Before leaving Dongola, Lord Wolseley had obtained information as to the state of supplies at Khartoum, which made it advisable to gain touch with General Gordon, who was shut up in that place, without delay. He determined therefore to arrange for the despatch of a mounted column across the desert to Metemmeh, where Gordon's steamers were awaiting them, while the remainder of the force was to follow the river in whalers. But food for the men and forage for the camels was lacking, and as this would not be forthcoming *en route* or at Metemmeh, supplies had to be collected before a start could be made, so that the column should carry its provisions with it. Camels and camel drivers were also scarce. Water was known to exist at the Gakdool (Jakdul) wells in quantity, and Jakdul was halfway on the journey.

The purchase of camels for riding and saddles was pushed on with speed. All available camel transport was ordered to the front. On arrival at Korti it was determined by Lord Wolseley that his force was to be divided, a part proceeding by water and the remainder by land. The land force, which was under the command of Brigadier-General Sir H. Stewart, was to consist of one and a half squadrons of the 19th Hussars, the four camel regiments, one battery of Royal Artillery, and a portion or perhaps the whole of the Royal Sussex Regiment. All the transport camels not detached for the water force were to accompany them. It was intended to garrison the Jakdul wells with the Royal Sussex Regiment and to collect there sixty days' provisions. The camel regiments were to march *via* Jakdul to Shendi.

On 30 December a message 'Khartoum all right' was received from Gordon. His messenger, however, gave verbal details of the exact position. Khartoum was besieged on three sides: Omdurman, Halfiyeh, and Hoggiali. Fighting was continuous day and night, but the enemy could not prevail otherwise than by starving out the garrison. Provisions were short. 'Come quickly, and come by Metemmeh or Berber. Send me word.'

Sir H. Stewart arrived at Korti on 15 December with the Guards and Mounted Infantry, Camel Regiments, and detachments of the South Staffordshire Regiment and Royal Engineers. Lord Wolseley and his personal staff and Sir C. Wilson arrived on the 16th, and Sir Redvers Buller with the remainder of the headquarters staff on the 24th.

On Christmas day three sections of the Light Camel Corps were at Korti, three were on their way from Debbeh to Korti, and three from Dongola to Debbeh. At Korti a number of camels promised by the Chief Saleh, of the Kababish tribe, were not to be found. Camel transport being therefore insufficient, it was impossible to move the whole force intended for the capture of Metemmeh across the desert in a body with their supplies, and also to form a depôt at Jakdul. It was therefore decided to send a convoy to Jakdul, there to form a post; the camels were then to return and bring on more men and supplies.

The strength of the Light Camel Regiment on this march was as follows: nine officers, eighty-one non-commissioned officers and men, and two hundred and fifty camels. The Heavy Camel Regiment had six officers, one hundred non-commissioned officers and men and two hundred and fifteen camels; the Guards Camel Regiment, nineteen officers three hundred and sixty-five non-commissioned officers and men and four hundred and six camels. The Heavy and Light Camel Regiments acted as transport, each camel carrying about 230 lb. and one man leading three camels. The column marched at 3 p.m. on 30 December, the order being as follows: Advanced guard, twenty men 19th Hussars, scouts of Guards Camel Regiment (2 scouts per company), not extended, Guards Camel Regiment. Royal Engineer Detachment, Moveable Field Hospital, half detachment bearer company, with all reserve water camels, baggage camels; 19th Hussars, Heavy Camel Regiment, Light Camel Regiment, Royal Artillery, Commissariat and Transport Corps, scouts of Mounted Infantry Camel Regiment, half detachment bearer company; six spare camels from Commissariat and Transport Corps, scouts of Mounted Infantry (two per company), not extended; rear guard, twelve men 19th Hussars. The distance between each of the corps was thirty yards.

The column marched till 7.30 a.m. the next morning, having only halted from 5 to 7.15 p.m. The distance covered was thirty-four miles. Short halts had been obligatory, however, to allow straggling camels to come up and to readjust loads.

The track across the desert was fairly good. The next day's march lasted from 3 p.m. to 8.30 p.m., when the wells of Hambok were reached, and where a little muddy water was found. At 1.15 a.m. on 1 January the force reached the well of El Howeiya and halted till 8.30 a.m. The water here was scanty and bad. Here a detachment of mounted infantry was left to improve the well. The column halted from 1 to 3.30 p.m. and then marched on till 7 p.m.

The moon rose at 8.30 p.m., when the column resumed its march and continued through the night. At 6.45 a.m. on the morning of 2 January they reached the gorge leading to Jakdul. At 2 a.m., before arriving here, the Abu Haifa wells were seized by a detachment of mounted infantry. The country now was thickly studded with mimosa and for the last twelve miles the road was flanked on the left by the escarpment of the Jebel Gilif range. The entire distance traversed from Korti to Jakdul was ninety-eight miles, and the time consumed sixty-three hours forty-five minutes, of which the march had occupied thirty-two and three-quarters hours. Very few Arabs were seen. Near Jakdul a party of men wearing the Mahdi's uniform was captured—they were on their way to Metemmeh.

The wells at Jakdul were found at a distance of two and a half miles north of the Korti-Metemmeh road. They were really pools or reservoirs rather than wells, as we understand the term, and they were three in number, besides some smaller pools. The best water, and fit for the men to drink, was found in the upper and middle pools; the lower pool was given over to the animals and for washing.

On the day he reached Jakdul Sir H. Stewart marched on his return to Korti with the whole of his force except the Guards Camel Corps and the Royal Engineers. These were left to garrison Jakdul wells and to improve the water supply. Pumps were rigged to bring water for the men from the middle pool and for the animals from the lower. Canvas troughs were set up for the animals, and a trench was also dug.

Sir H. Stewart reached Korti at noon on 5 January, leaving a small detachment at Hambok on irrigation duty.

There was no sickness among the men on these two marches, but thirty-one camels out of two thousand one hundred and ninety-five died and many more were incapacitated for further work.

On 7 January Colonel Stanley Clarke with three sections of the Light Camel Regiment escorted one thousand camels from Korti to Jakdul: one hundred of these carried small-arm ammunition, eighty medical stores, thirty artillery stores, and the remainder supplies. Arrived at Jakdul, Colonel Clarke returned with the unloaded camels to Korti. On 8 January the main body of the desert column again left for Jakdul. The Mounted Infantry at El Howeiya was relieved by fifty men of the Essex Regiment.

A message from Lord Wolseley to Major-General Buller, dated Korti, 5 February 1885, contains the following passage concerning the Light Camel Regiment:

> To strengthen you in camel troops I am sending you as many of the Light Camel Regiment as I can mount efficiently. They will leave Korti this evening, and should be with you on the 8th instant.

Much had happened. Khartoum had fallen; Gordon had been slain. The expedition which made its dash by water to relieve him had arrived too late to effect its purpose. The steamers which conveyed the troops had in returning met with disaster, being wrecked upon sunken rocks, and Sir Charles Wilson and his party who had been cast away when the *Bordein* was wrecked, had been rescued by Lord Charles Beresford on the *Safieh* and brought back to Gubat. The situation was changed, The death of Gordon had, in the opinion of the Government, rendered any further advance unnecessary. A general halt took place, and preparations far the evacuation of the country up the Nile were, if not actually ordered, certainly imminent.

From Lord Wolsely came directions to General Boiler that if he should find himself temporarily cut off from Korti, he was to take steps to evacuate Gubat and to concentrate all the troops that had crossed the desert to Jakdul, on the wells of Abu Halfa and Howeiya. If the evacuation of Gubat, owing to the movements of the enemy, should prove difficult, the garrison of that place should proceed to Jakdul and even that of Abu Klea. A siege of Gubat by the enemy, whose troops were now largely increased by the release of forces from Khartoum, was much to be deprecated.

Lord Wolsely was prepared to 'advance on Khartoum' and 'defeat the Mahdi,' but as he telegraphed home, 'operation under present conditions is much more difficult than before, and owing to the lateness of the season, would be somewhat hazardous, for our serious enemy would be the hot weather, not Mahdi.' He estimated that he could not possibly reach Khartoum 'with sufficient force to attack Mahdi for six weeks' from the date of writing, 4 February. It must be remembered that though there were weighty reasons for believing in the death of Gordon, there had not as yet been absolute proof of that unhappy event. A telegraphic reply was received on 7 February, in which the decision was practically left to him by the Government.

On 8 February Lord Wolseley telegraphed expressing a slight hope that Gordon might still be holding out at Khartoum, but that unless this was the case he could not do more than capture Berber that season, as reinforcements could not now reach him in time for the present season's campaign on the Nile. General Buller was now ordered (10 February) to take Metemmeh as soon as he should find himself strong enough, and also to use his discretion as to the occupation of Shendi. When Metemmeh was taken he was to combine with General Earle and attack Berber. These orders reached General Buller on 13 February, when he had already partly evacuated Gubat, and was leaving that place himself on the morrow.

General Buller wrote that he thought it advisable to adhere to his former opinion and to evacuate Gubat and occupy the Abu Klea wells without attacking Metemmeh, his principal reason for this being that he was convinced that a force was *en route* for Khartoum, and that if this force took up a position in the bush between Gubat and Abu Klea his communications could with great difficulty be kept open; whereas at Abu Klea his men would be secure, and moreover no force of the enemy could exist in the desert between that place and Jakdul.

On 12 February Lord Wolseley received news of the successful action of the river column at Kirbekan and of the death of General Earle. On the morning of 13 February a convoy was despatched by General Buller to Korti under command of Colonel Talbot. All the sick and wounded, including Sir H. Stewart, accompanied it. The sick and wounded, who numbered seventy-five, were in the charge of the Bearer Company under Surgeon-Major Conolly, those unable to walk being carried on stretchers by Egyptian soldiers. Three hundred Egyptian soldiers and camp followers as well as a part of the Commissariat and Transport Corps also accompanied the convoy. The escort consisted of a wing of the Heavy Camel Regiment, the Marine Company and another of the Guards Camel Regiment, and a company of the Mounted Infantry Regiment.

After a march of eight or nine miles the convoy halted for breakfast in the bushy district of Shebakat. As soon as the march was resumed the scouts reported a large convoy of camels with Arab drivers in sight. Skirmishers were sent out to reconnoitre, and if possible to capture the convoy of the enemy. The enemy, however, proved too strong to be tracked, and the skirmishers were then reinforced. Colonel Talbot then drew up his force in the following manner: The men of the Heavy Camel Regiment were extended to the right front, the Mounted Infantry in rear of the column. The Egyptian soldiers were posted in line on both flanks, while the sick and wounded were placed in the middle of the camels. For an hour and a half the enemy kept up a fairly well directed fire, and from three sides of the convoy.

A body of troops was now seen advancing on the left flank of the convoy, and was received by a volley. Fortunately the shooting of the Egyptians was not good, for the advancing troops turned out to be the Light Camel Regiment, which, under command of Colonel Stanley Clarke, had arrived on the scene from Jakdul. At 1.15 the enemy fired a parting volley and withdrew. The convoy, accompanied by the light Camel Regiment, then proceeded on its way to Abu Klea, where it duly arrived on the morning of the 14th and halted till 8.30 on the morning of the 16th. In this brush with the enemy the casualties were two killed and six wounded. On 17 February Sir H. Stewart died at a spot about seven miles north of Jebel El Nus. He was buried next day at Jakdul. On arrival at Jakdul the convoy found General Buller absent and Sir Evelyn Wood in command.

On 14 February the evacuation of Gubat began. The force then numbered one thousand seven hundred. Both officers and men were all on foot except the 19th Hussars. One camel was allotted to every four men.

The force in its retreat was followed by a small body of the enemy's cavalry, but at a respectful distance, a fact which was much to the disappointment of the troops, who were most eager for a brush with the enemy.

The wells at Abu Klea were reached at 11 a.m. on 15 February. Here the Light Camel Regiment was found, an addition to the garrison which Lord Wolseley had sent to relieve the Heavy Camel Regiment. At Abu Klea Buller halted, but finding the water supply and food for the horses and camels likewise insufficient, despatched the 19th Hussars, except one officer, six men, and eight horses, the Guards Camel Regiment, the remainder of the Heavy Camel Regiment, the Transport Corps, and all spare camels, together, with all the Sudanese in camp, to Jakdul. Still there was not food enough for the camels remaining with him, and consequently later in the day he sent one hundred of the Light Camel Regiment with one hundred and fifty camels to Jakdul under Lieut.-Colonel J. P. Brabazon.

On the evening of the 16th the enemy were found to have occupied a hill commanding the camp and distant there-from some one thousand two hundred yards to the north-east. From this point during the night they kept up a harassing fire. They were driven away in the morning greatly owing to the action of Major F. M. Wardrop, D.A.A.G., who, with three other mounted men and Lieutenant R. J. Tudway, Mounted Infantry, 'by appearing rapidly and firing from several points in succession made the enemy believe that their position (which was a very strong one) was threatened in the rear, and caused them to evacuate it.'

On the afternoon of 17 February Lieut.-Colonel Hugh McCalmont, Light Camel Regiment, arrived with a despatch from the Chief of the Staff which brought news of the successful action at Kirbekan. All was quiet that night. In the fighting of the 16th and 17th three men were killed, four officers, among whom was numbered Captain Harold Paget of the Light Camel Regiment; and twenty-three men were wounded.

On 20 February we find camels for mounting the troops greatly needed. The Heavy Camel Regiment and the Guards Camel Regiment required nearly all new ones; the Light Camel Regiment at least one hundred, the Mounted Infantry two hundred, and all the transport.

On 23 February Abu Klea was evacuated. About 11 a.m. on that day a reinforcement of some eight thousand men reached the enemy. General Buller then filled in the larger wells and abandoned the forts. It had been intended to destroy the forts and spare the wells. At 2.p.m. all the baggage with an escort of three hundred men under Colonel Stanley Clarke. marched out with orders to camp on the Oh Mit Handel plain, out of gunshot range of the Abu Klea hills. At 6.40 p.m. the garrisons of the outposts were withdrawn, and at 7.40 the whole force marched unmolested out of Abu Klea. At noon on the 24th some of the enemy's scouts appeared and opened fire, but shortly retired. All men were on foot. The allowance of water was three quarts *per diem*. The sick and wounded numbered thirty-two.

THE LIGHT CAMEL CORPS: 1884–1885

Buller's force reached Jakdul about noon on 26 February 1885. The withdrawal of troops then proceeded gradually. By 16 March the last had arrived at Korti from the desert. During the summer the Light Camel Regiment at Sbabadool, a place forty miles from Dongola. Colonel Stanley Clarke had now returned home, and Colonel Hugh McCalmont commanded the Regiment.

From the following account, which has kindly been furnished by Major-General Sir Hugh McCalmont, K.C.B., C.V.O. Colonel of the 7th Hussars, we have obtained in brief narrative form a consecutive story of the proceedings of the Regiment in Egypt:

On arrival at Cairo, the Regiment was marched out to the Pyramids into camp, but was almost immediately brought back and entrained for Assouan, where camels had been collected and whence it marched up the Nile to Korti. The detachment to which the officers and men of the 7th were posted left Assouan, on the second day after their arrival at that place. It was probably about fourteen days before the last detachment of the Light Camel Regiment was able to leave Assouan. The Light Camel Regiment was not present at either of the engagements at Abu Klea or Gubat. The whole of the Camels of the Camel Corps had to be used for filling up Jakdul, some half-way across the Bayuda Desert towards Metemmeh, and the Regiment was employed on convoy duty. Immediately on receipt of the news of the fall of Khartoum, Lord Wolsely sent the Light Camel Regiment across the Desert to Gubat, whence General Boner retired without attempting the capture of Metemmeh. While on the retirement from Gubat to the Wells of Abu Klea the Regiment acted as Rear Guard to the Column composed of the three other Regiments of the Camel Corps, the Field Artillery and one Squadron of the 19th Hussars. On this march some desultory skirmishing took place. The sick and other impedimenta were sent back to Korti, and the force lay entrenched at Abu Klea for several days. Until certain points were occupied by works some firing took place from the positions occupied by the enemy, who were distant about 1200 yards.

While at Abu Klea several casualties occurred, among those

wounded being Captain (now Colonel) Harold Paget, whose injury was severe. The 1st Battalion 18th Royal Irish had also arrived at Abu Klea when the force reached that place. The return march, after passing some days at Abu Klea, was trying. The Rear Guard, furnished by the Royal Irish, was in touch with the enemy for the first thirty miles. Water was very scarce and all that the force possessed required to be carried.

Soon after the column arrived at Korti, Colonel Stanley Clarke (the late Major-General Sir Stanley de Astel Calvert Clarke, G.C.V.O., C.M.G.), who was Brigadier of the Camel Corps, returned home, and the command devolved upon Lieut.-Colonel Hugh McCalmont, who had already commanded it on the march from Abu Klea.

The Regiment subsequently went into a sort of hut encampment under the palm trees at Shabadad on the Nile until the evacuation of the Province of Dongola in June 1885. Captain T. H. Phipps, 7th (Queen's Own) Hussars, who was second in command of the Mounted Infantry Camel Regiment, died on his way home, at Cairo. Two officers of the Light Camel Regiment died of fever at Shabadad, and several men, while nine officers and a number of men had been invalided home before the return of the Regiment. The Light Camel Regiment consisted of nine troops. The officers, non-commissioned officers and men were selected from the Hussar regiments at home.

The names of the officers selected from the 7th Hussars have already been given. The service therefore, as will be seen, was hard and trying, without the compensation of being engaged in the battle at Abu Klea. For the greater part of the time the Regiment was about 1000 miles from the rail head. Rations for months were none too plenty owing to the difficulties of transport, and, thanks to the climate, by no means super-excellent in quality or in quantity.'

On 1 June 1885 they started down the Nile, embarked and arrived in England on 23 July.

Four officers of the 7th Hussars were named whose services were deserving of special mention in Lord Wolseley's final Despatch, dated 15 July 1885: Brevet Lieutenant-Colonel H. McCalmont, Captain C. F. Thomson, Captain T. H. Phipps (since dead), and Captain Harold Paget.

On arrival in England the Regiment, being at Aldershot, was joined there by the detachment which had been absent in the Soudan.

In October the quarters were shifted from Aldershot to Hounslow, Hampton Court, and Kensington, E Troop being at Hampton Court and G Troop at Kensington.

Only two entries regarding 1886 need be here entered: the Cambridge Challenge Shield was won by the Regimental Team. The Regiment left Hounslow on 21 September and marched *via* Croydon to Shorncliffe to prepare for embarkation for India.

CHAPTER 8

East Africa
1886

On the receipt of orders for the Regiment to proceed to India preparations began apace. Arms and accoutrements were overhauled and horses were handed over to other regiments. One hundred went to the Mounted Infantry, but the majority were held back until the arrival of the 14th Hussars from India, when they received them after landing and reaching their station. Men from the 10th, 14th, 19th, and 20th Hussars were also transferred to the 7th.

On 25 November 1886, twenty-one officers, five hundred and eighty-seven non-commissioned officers and men, fifty women, and forty-seven children proceeded from Shorncliffe *to* Portsmouth by rail and there embarked on H.M. troopship *Euphrates*. The trooper sailed on the morrow, and many friends of the Regiment journeyed to Portsmouth to give them a hearty send-off. Colonel Drew was in command. The voyage was prosperous and the Regiment arrived at Bombay on December 23. Having disembarked, the 7th proceeded by train to Poona, where they spent their Christmas amid their new surroundings.

On 26 December the Regiment was inspected by H.R.H. the Duke of Connaught, Commander-in-Chief of the Bombay Army. Next day the 7th entrained for Wadi, where they rested for a few hours, and then proceeded to their destination Secunderabad, where they arrived on the morning of 28 December. The depôt of the Regiment at home was at Canterbury.

On 19 February 1887 the Regiment was inspected by Major-General W. A. Gib, C.B.

As Colonel Drew's period of command had now expired he was succeeded by Colonel A. Peel on 31 July.

During 1887, the first year of the stay of the Regiment in India, there were many cases of enteric fever, and fifteen cases unfortunately proved fatal: Lieutenant H. L. Warren, twelve men, and two women died.

The information recorded of the years 1888-9 is very scanty.

In January and February the Regiment was inspected by Brigadier-General Luck, C.B., and Major-General Gib, C.B. The first-named officer also inspected them a second time on 31 October.

During 1889 they were inspected by Brigadier-General H. Bengough, C.B., on 3 February, and by Brigadier-General Luck, C.B., on 4 November and following days.

The period of command of Colonel Peel had now expired, and he was succeeded by Lieutenant-Colonel John L. Hunt on 18 December

During 1890 we find the Regiment taking part in a camp of exercise from 20 January to 3 February.

On 10 and 11 February they were inspected by Major-General C. J. East, C.B.

No events are recorded until 6 November, when the right wing went out to camp.

The camps of the two squadrons were situated about ten or twelve miles apart and the squadrons were exercised in reconnaissance and outpost duties against one another for seven days. At the end of the month the left wing went into camp and was exercised in a precisely similar way. These camps of exercise were apparently intended to act as a kind of preparation for the more extended manoeuvres which were held during January 1891.

On 2 January the Regiment went out to a camp of exercise. A tract of country extending for about forty miles westward of Secunderabad was selected for manoeuvres, and until 10 January the force, which was divided into two mixed brigades, was employed in manoeuvring one against the other. On that date the whole division went into a standing camp near Nizampett under the command of Major-General C. J. East, C.B., who commanded the district. His Excellency the Commander-in-Chief of Madras was present in camp. The Cavalry Division consisted of five cavalry regiments and

1886

a battery of the Royal Horse Artillery. General Luck was also at the camp and the Cavalry Division was exercised under his command.

The inspection of the Regiment by the Inspector-General of Cavalry began during the stay of the 7th in the standing camp and concluded in cantonments on 22 January.

On 30 and 31 January 1891 the Regiment was inspected by Major-General C. J. East, C.B.

During the month of May a report was received on the horses of the 7th Hussars, dated 23 April. The inspection was held by the Inspecting Veterinary Surgeon of the Madras Army at Secunderabad. The following is extracted there-from:

I have never seen in the Army, better groomed horses, with greater bloom on their coats, and all the signs of health, and just in the proper condition.

31 August, 1891. His Excellency Sir James Dormer, Commander-in-Chief, Madras Army, arrived at Secunderabad. He inspected the barracks and institutions of the Regiment on 2 September.

On the same day the 7th Hussars were exercised under Major-General East, C.B., in reconnoitring, and in the attack of a position. The defending force was under the command of Brigadier-General Protheroe, C.B., C.S.I., Commanding the Hyderabad contingent. Prior to returning to Ootacamund on 5 September, the Commander-in-Chief dined on the previous day with Colonel Hunt and the officers of the Regiment.

In October the 7th Hussars proceeded to Mhow by rail to relieve the 18th Hussars. The Regiment travelled thus: 2nd Squadron under Major Reid on 2 October. 4th Squadron and married families under Major Thomson on 22 October. 1st Squadron and Chargers under Captain Nicholson on 23 October. 3rd Squadron and Headquarters under Lieut.-Colonel Hunt on 24 October.

17 November 1891 the Regiment was inspected by his Excellency the Commander-in-Chief of the Bombay Army, Lieut.-General Sir George Richards Greaves, K.C.B., K.C.M.G., who visited the station. It would appear that inspections were exceedingly elaborate at this period ; the dates are as follows: Brigadier-General G. Luck, C.B., 29, 30, and 31 December 1891; Major-General H. S. Anderson, C.B., 25, 26, 27, and 29 February 1892. By the same officer on 27 January, 1 February, and 3 March 1893. By the

Commander-in-Chief of the Bombay Army, Lieut.-General Sir G. Greaves, on 4 February and by Major-General G. Luck on 27, 28, and 29 February. Thus between 27 January and 3 March 1893 no fewer than seven days out of thirty-six were so employed.

On 2 June a letter from the Adjutant-General in India was received which enclosed a copy of a Horse Guards letter, dated 2 May 1893 conveying Her Majesty's approval of the word *Orthes* being borne on the appointments of the 7th (Queen's Own) Hussars in commemoration of the part taken by the Regiment in the battle fought there on 27 February 1814.

On 25 June 1893 a very sad accident occurred at Poona by which the lives of three officers of the Regiment were unhappily lost. It appears that Lieutenants Crawley and Sutton and Second Lieutenant the Hon. H. P. Verney, the second son of Lord Willoughby de Broke, went down to the Poona Boat Club about 3 p.m. and engaged a sailing boat called the *Una* which belonged to the Connaught Boat Club at Kirkee, but which was lying at the other club's boat-house. The party set sail, and according to the statement of the boatman were about to tack across when they were suddenly carried into mid-stream, where both the wind and current were against them. Despite all their efforts to tack or to bring the boat up, they evidently lost all control over her, and she drifted rapidly down mid-stream towards the Bund below. The unfortunate officers made desperate endeavours to reach the opposite bank but the boat kept on her course, and as she was approaching the Bund they lowered the sail suddenly, and when the boat was whirled over, according to an eye-witness they jumped with it. Down went the boat with a crash on the rocks below and appeared there lying on her side with the three men clinging to her. With the fearful torrent from above and the whirling waters all around them they clung for a few minutes, and then the boat and the unfortunate officers disappeared under the waves. Two of them never rose again, but the third was seen struggling in the wild waters, and had nearly reached the centre arch of the Bund bridge when he too disappeared. The body of Second Lieutenant the Hon. H. P. Verney was recovered on the 27th, those of the other two officers on the 28th. The funeral of all three officers took place with military honours at 5 p.m. on that day, the band and firing party being furnished by the

2nd Battalion Yorkshire Light Infantry; the gun-carriages by the L Battery R.H.A. The colonel, the major, five captains and three lieutenants of the 7th Hussars as well as the regimental sergeant-major, a squadron sergeant-major and a sergeant came to Poona for the funeral. At the time of the accident Lieutenant Sutton was in Poona to attend a signalling class, while the other two officers were there on leave.

On 20, 21, 23, and 24 October, Major-General Luck, Inspector-General of Cavalry, inspected the Regiment.

The period of Lieut.-Colonel J. L. Hunt's command having expired on 18 November, the command of the Regiment was taken over on 19 November by Lieut.-Colonel Harrie Archbold Reid.

The events recorded for the year 1894 are but two in number. On 15 and 16 January the 7th Hussars were inspected by his Excellency Lieut.-General C. E. Nairn, Commander-in-Chief, Bombay Army, and the Regiment subsequently took part in the camp of exercise held in the neighbourhood of Mhow under the direction of Major-General Anderson, C.B.

On 26 June Lieut.-Colonel H. A. Reid having retired on half-pay owing to ill-health, Lieut.-Colonel Harold Paget was promoted to the command of the Regiment.

In the Regimental Record there is no entry telling us that the Regiment was about to leave India and to proceed to Natal. Whether this change came about in the ordinary course or whether the move was a sudden one is not stated. We merely have under date 10 October 1895:

> The Regiment embarked at Bombay on the hired transport *Victoria* and arrived at Durban, Port Natal, on 22 October.

It would appear from information subsequently received from Colonel Harold Paget that the move of the Regiment from Mhow to Natal was made in the ordinary course of reliefs, the 7th Hussars taking the place of the 3rd Dragoon Guards in Natal. The Regiment travelled by train to Bombay with its women and children, handing over its horses at Mhow to the 20th Hussars, who had arrived from England a few days before the departure of the 7th Hussars.

The horses of the 3rd Dragoon Guards were taken over in Natal. The Regiment proceeded from Durban to Pieter Maritzburg by train. And now, after a peace service winch had extended over a period of eleven years, the 7th Hussars were again about to be actively employed. A serious condition of unrest had for some time manifested itself among the natives of Matabeleland and Mashonaland. This unhappy state of affairs had culminated in a series of atrocious murders, the looting of stores, and all the circumstances which accompany the existence of a widespread rebellion.

Hence it was that about nine months later the Regiment again returned to Durban, marching thither from Pieter Maritzburg, and under orders to embark for East London *en route* for Mafeking, at which place the British troops were being assembled to suppress the revolt.

The three squadrons accordingly embarked at Durban, the *Goth* having been brought inside the harbour for the purpose. In reply to a question as to the general health of the horses and the losses during the subsequent campaigns Colonel Paget stated as follows:

The loss of horses was certainly heavy, as it always is in any campaign, although most of the animals had been some time in the country. There was not any very severe epidemic of what is known as horse sickness in South Africa, but there was great wastage from the unavoidable effects of marching with occasional short rations. The losses of the two squadrons that were engaged in Matabeleland were made good in that country, and finally at the conclusion of the operations the horses were handed over to the authorities in Matabeleland and the two squadrons returned dismounted to Natal, where the Regiment was practically remounted the following year by importations from Argentina.

We will now quote the *Manuscript Regimental Record* for the year 1896 from 1 May to 28 November:

1st May.—Three Squadrons of the Regiment marched from Pieter Maritzburg and embarked on s.s. *Goth* for East London to join the force concentrating at Mafeking for the suppression of the native rising in Matabeleland. Mafeking was reached on May 8th, 9th, and 10th.
20 June.—A Squadron under Major Carew marched to Macloutsie, followed on 25 June by D Squadron under Captain Agnew.

25 July.—The force moved on to Tuli under command of Lieut.-Colonel Paget

18 August.—The force arrived at Fort Victoria.

24 August.—In accordance with orders received from Major-Genera Sir F. Carrington, Commanding the Imperial Troops in Rhodesia, column started for Gwelo, which was reached on Sept. 4, Monogola's stronghold being taken on the way with slight loss.

7 September.—The force marched into the Kevekwe District and returned to Gwelo on the 23rd.

October.—After destroying 'Ndema's, Monogola's, and Wedza's strongholds the column marched to Buluwayo, which was reached on 28 *Nov.*, when the Regiment went into standing camp.

Thanks, however, to the courtesy of Major C. Norton, who was at the time serving in the Regiment, and is now second in command, we are enabled to give a full and detailed account of these operations. In so using the mass of information, some sixty-eight pages of foolscap, we shall not here quote *in extenso* but shall reduce the Diary to which we have access into narrative form.

On 13 April 1896 orders were received at Pieter Maritzburg from the High Commissioner through the Chief Staff Officer at Cape Town for three squadrons of the Regiment to proceed to Mafeking in order to suppress the Matabele rising. On 21 April this movement was suspended by orders from home. The reason for this change was, that four hundred Mounted Infantry in three companies were to be sent from home to the Cape. Hence the movement of the 7th Hussars from Natal was suspended, but the Mounted Infantry already in Natal, which had been ordered up country, were to proceed as already arranged for.

On 22 April this counter order of the 7th Hussars was suspended. Next day the 7th Hussars were ordered to go 'as previously arranged.' The officers of the Regiment who were engaged on this service were as follows: Lieut.-Colonel Harold Paget, Commanding the Imperial Contingent; Major H. M. Ridley; Captains Carew, Agnew, and FitzHenry; Lieutenants Poore, Dalgety, Vaughan, and Wormald; 2nd Lieutenants Rankin, Imbert-Terry, Greville, Holford, H.S.H. Prince Alexander of Teck, and Rawstorne; Lieutenant and Adjutant Norton and Quartermaster Coe. The other units of the force included Mounted Infantry (2nd West Riding Regiment), the 2nd York and Lancaster Regiment, Army Medical Staff, Army Veterinary Department, and Army Pay Department.

On 27 April nineteen non-commissioned officers and men under Lieutenant Vaughan left Pieter Maritzburg for Durban by train to embark on the s.s. *Lismore Castle* for East London *en route* for Mafeking to act as an escort to a section of the 10th Mountain Battery, Royal Artillery, which left by the same train under Lieutenant McCulloch, R. A., Lieutenant Poore also went at the same time to make arrangements for the disembarkation of the force at East London.

1 May, C and D Squadrons of the 7th Hussars and three sections of the Mounted Infantry marched at 10 a.m. from Pieter Maritzburg, camping for the night at Half Way House near Botha's Hill, a distance of about twenty-eight miles. A squadron left Pieter Maritzburg by rail at 11 p.m. Apparently the whole squadron did not go, as on May 2—

> 2 sections M. I. and the remainder of A squadron 7th Hussars marched at 6.30 a.m. C and D squadrons having arrived marched on 2 May, the latter at 7.30 a.m. and the former at 8.15. The first party reached Durban Point, a distance of twenty-seven miles, at 1.45 p.m.

A Squadron and a section of Mounted Infantry from Eshowe had already embarked on the *Goth*. The remainder of the force at once began to embark on arrival, but operations had to be suspended at 6 p.m. owing to the inadequate lighting of the ship. On the march the Regiment lost five horses: four actually died and one was left at Durban unfit to proceed.

On Sunday morning, 3 May the embarkation of the horses began again and was completed by 1.45 p.m. One horse was accidentally killed on board ship by timber falling through a hatchway. At 3 p.m. The *Goth* sailed with twenty-two officers, four hundred and fifty-six non-commissioned officers and men, fifty-five officers' chargers, and four hundred and thirty-three troop-horses.

The *Goth* anchored off East London about 8.15 a.m. on 4 May. Two hours later the disembarkation of the horses began and continued till 6 p.m. A squadron was the first to be set on shore, and this was done as completely as possible, but as stores could not be landed, it was not permissible for them to start at once. By 6 p.m. two hundred and eighteen horses had been placed in lighters without any mishap; by 10 p.m. the lighters had transferred their lading to shore. All this would have occupied much less time had there been a sufficient number of lighters; unfortunately there were not. On Tuesday, 5 May, at 8 a.m. the transfer of the remaining two hundred and seventy-one horses to the lighters was resumed. By noon one hundred had been placed thereon. The disembarkation of the men followed in a tug, the *Midge*. By 5 p.m. only seventeen horses and eighteen men of the 7th Hussars remained on board; of the Mounted Infantry forty-five horses and eighty-one men. The disembarkation was resumed on the morrow at 7 a.m. and was completed by 1 p.m. without any casualties.

The men and horses were conveyed to Mafeking by train as follows:

5 May.—First train, A Squadron with horses. Second train, C Squadron with horses and accompanied by Lieut.-Colonel Paget and Lieutenant and Quartermaster Coe.

6 May.—First train, D Squadron with horses and Major Ridley; second train, Mounted Infantry and forty men and horses of the 7th Hussars; third train, Mounted Infantry and eight men and horses of the 7th Hussars, accompanied by Lieutenant and Adjutant Norton. Between 7 a.m. on 8 May and 3 a.m. on 10 May the various parties reached Mafeking without any casualty to either man or horse. Here Lieut.-Colonel Paget assumed the command of the Imperial Contingent, Matabeleland Relief Force. It being Sunday a church parade was held at 9.30 a.m. for all troops in camp.

Monday, 11 May.—Owing to the heavy rain it was impossible to hold any parades. The next two days were occupied in exercising the squadron under their own commanders and an inspection of horses. On 13 and 14 May the guns left for Bulawayo for service under the British South Africa Company. On this same day two horses of A Squadron died. Parades and field exercises now took place daily, varied only by inspections of saddlery, kits, and equipments, church parades, clothes washing in the river, and instructions in wagon-packing.

On 20 May the defence of a laager was practised under Lieut.-

Colonel Paget. Next day Sir Frederick Carrington arrived from Cape Town accompanied by his Brigade Major, Captain Vyvyan (East Kent Regiment), and his A.D.C. Lieutenant Ferguson (South Wales Borderers).

22 May.—Lieut.-Colonel R. S. Baden-Powell (13th Hussars), Chief Staff Officer to Sir F. Carrington, arrived. The Commander of the Forces left for Bulawayo with his staff at 2 p.m. On 23 May by coach.

25 May, being Whit Monday, was kept as a holiday as far as possible.

26 May, being observed as the Queen's Birthday, a full parade was held on the polo grand, with a Royal Salute, cheers for Her Majesty, and a march past. The 7th Hussars among other things galloped past in line, did the Sword Exercise and Pursuing Practice, and the function concluded with an advance in line at a trot of the whole force.

From 27 May until Saturday, 20 June, parades, exercises, and washing days are all that have to be recorded, except that on Saturday, 6 June, Private Gould of the Regiment was declared a deserter since May 14, and on 8 June three Boer officers visited the camp officially. They were entertained by Lieut.-Colonel Paget, and the object of their visit was to ascertain if there was any truth in a rumour current in Pretoria that 'British troops were massing here.'

On 11 June Lieutenant Imbert-Terry's second charger was found to be suffering from glanders and destroyed. On 16 June there was an alarm at 2 p.m. and the whole force turned out mounted by 2.15. Two privates were this day absent without leave. Next day two more were missing from D Squadron, and piquets were sent into the town and patrols along the Bulawayo and Johannesburg roads to find them.

On Thursday, 18 June, news of the rising of the natives in the Salisbury District on the previous day arrived. During this time the squadron training had been in progress. It was completed by 19 June, when the inspection of C Squadron began and was continued on the following day. On the same date a telegram was received from the D. A. A. G,. Bulawayo, asking for five hundred Imperial troops to be moved to Macloutsie, subject of course to approval from Cape Town, The confirmation from the High Commissioner arrived a few hours after, by which orders ere received for the first party to prepare to move at once. Later the request for five hundred men was reduced to two hundred. This force was to be thus composed: one hundred of the 7th Hussars and a like number of Mounted Infantry, with fifty spare horses for casualties. Accordingly the first party immediately entrained for Macloutsie. The 7th Hussars (part of A Squadron) were under the command of Lieutenant Vaughan and Second Lieutenant Greville. They started at 7.30 a.m. on the 22nd, and were to proceed by rail as far as Lobatsi and the remainder of the journey by route march.

The second party left at 8 a.m. on 23 June under the command of Captain Carew and Second Lieutenant Holford. The third and fourth parties left similarly on the 24th and 25th. Hardly had the last train started for a couple of hours when a wire arrived stating that the position in Mashonaland required an additional two hundred cavalry, and begging that they might be sent at once to Macloutsie. Supplies for two hundred men 'as 6 weeks for 400' were to be sent to meet this new force at Macloutsie by the Administration. This force was to be composed as the former two hundred.

On 26 June Captain Carew's party reached Gaberones, but 'mules and wagons very bad, had difficulty also with conductor who refused to go on,' was reported by that officer. That day Lieutenant Vaughan's party left Gaberones. One of Captain Carew's wagons broke down on the road and had to be left behind. The fifth marching party had been intended to start on 30 June, but its departure was postponed for a day, owing to the inadequate supply of grain at Gaberones. It accordingly did not entrain till Wednesday morning, 1 July. The Hussars were under the command of Lieutenant R. M. Poore and Second Lieutenant Imbert-Terry. Meanwhile the Mounted Infantry were having trouble on the road, the pace of the leading party being less than that of the one which followed.

Lieutenant Vaughan reached Palla at 9 a.m. on 1 July. He reported that one of the men, by name Peters, had been absent at Gaberones and was left behind. He also asked for an additional wagon, and this was bought for him. The fifth marching party left Lobatsi on 1 July. The sixth, under the command of Captain Agnew and Second Lieutenant H.S.H. Prince Alexander of Teck, started on 2 July. Lieutenant Vaughan was now ordered to remain at Palla to await the arrival of Captain Carew's party and then for both to proceed together, sharing the extra wagon.

A seventh party left for Lobatsi on 3 July; this was composed mainly of officers and men of the 2nd York and Lancaster Regiment. Captain Carew now found that the grain requisite for the journey between Palla and Macloutsie could not be carried with only one extra wagon, and that the whole squadron could not proceed together owing to the insufficient supply of water. He was informed, in reply to a wire to this effect, that the squadron must proceed together, but all reserve ration wagons were to be left behind to follow on the next day.

An eighth party, also composed of the 2nd York and Lancaster Regiment, left on 4 July after some delay caused by the derailment of two of the horse-trucks. During the next three days there is no event to record. On 8 July Colonel Paget and Lieutenant Norton left for Macloutsie by train at 7 a.m. They reached Aasvogel kop at 5 p.m. and there took coach, starting at 6.30 p.m. They reached Gaberones at 7a.m. and left again two hours later, having found the eighth marching party already arrived. On the morrow they reached Mochudi, and Palla on the 11th. Here Lieutenant Poore's party was expected. Colonel Paget left again at 9.30 a.m. and arrived at Palapye at 7 a.m. On 13 July. An hour's delay took place owing to a wheel of the coach coming off. From Palapye they started in a special Cape cart drawn by four mules at 3 p.m., leaving Corporal Paterson, the servants, and most of the baggage to follow as soon as possible. On the road the driver lost his way in the dark and they had to outspan at 6 p.m. And wait for daylight. The journey was resumed at 7 a.m. next morning and continued till 11 p.m. On the 15th they inspanned and started at 6 a.m. and halted from 5 p.m. to midnight to rest the mules. Proceeding again on their journey the cart upset in a river at 3 a.m. and the travellers were obliged to make a fire and wait for daylight. However, at length, on 16 July, at 7.30 a.m., they reached their destination, Macloutsie, where they found the A Squadron had arrived all well on the previous day.

On 17 July, having telegraphed for instructions, a reply was received ordering the first party to march to Tuli, there to pick up twenty wagons from the Transvaal which were due there on the 20th. These wagons contained stores which were to be taken on to Victoria partly for their own use and partly for Victoria. Colonel Paget was ordered on arrival at that place to assume command of the Victoria District. He was to consult the Civil Commissioner as to what natives should be cleared out. Chaka was named as a possible one. He was also informed that a strong column was then operating from Victoria against Indema and was among other things ordered to keep a sharp look-out for fugitive *impis* retiring from Matoppos towards Gayaland, South-East.

Some of the natives around Victoria were reported to be friendly; maps were promised, and it was stated that the second marching party would probably proceed to Bulawayo. The chief staff officer at Bulawayo now telegraphed that the General Officer Commanding wished the column to take tents and, in consequence, it would be needful to wait probably for a week until the eighth marching party arrived, as all the tents were with them.

The third and fourth marching parties arrived on Saturday, 18 July. There had been no casualties among the men, but there had been great trouble with the transport, many mules being in very poor condition and several of the native drivers refusing to go on. Next day the disaffected drivers were discharged and steps were taken to replace them. Forage was evidently scarce, as the mules of the first and second marching parties had to be sent out about thirteen miles to graze.

After a delay of two days the order to take tents was cancelled, and they were ordered to make their own arrangements for having them sent on after them. But the Maxims had not arrived. They had, however, left Palapye and might be expected to reach Macloutsie on the 24th. However, the mules that had been sent out to graze were brought in.

On 21 July news came by wire from Bulawayo that Babyan's impi had been defeated, sixty Matabele had been killed and five white men. From Victoria also came news that the Chief Setoutsi was probably unfriendly, and that the road from Tuli to Mtipis was 'considered risky.' Twelve native scouts would be awaiting them at Tuli. It was now decided to send the Mounted Infantry detachment out to a spot about seven miles along the Tuli road where the grazing was good. They were to take eleven wagons with them and a water-cart.

It was now ascertained that the natives in the direction of Mtipis were quiet and peacefully inclined, but nothing came to hand regarding the Chief Setoutsi.

Meanwhile the mules in the Maxims were in a very bad state, and Dalgety, with whose party they were, reported by wire that it would probably take him six days more to reach Macloutsie. This was on 22 July, and the Maxims had been expected on the 25th. Arrangements were then made to send a wagon out 20 miles to meet Dalgety, but as his precise road was not ascertained, nothing could be done. About 2 p.m. on that day the Mounted Infantry under Major Rivett-Carnac left with eleven wagons and a water-cart, but at the first drift the leading wagon broke its disselboom, and owing to the delay they were compelled to outspan after proceeding only three miles. However, Lieutenant Poore arrived with the D Squadron (first detachment).

A proclamation was now issued by General Goodenough for the benefit of the rebels, in which their position was stated in precise terms, and setting forth the pains and penalties to which they were subject for various acts and crimes.

On 23 July news came from Victoria that Setoutsi had cleared out of the district and that the road was now safe. Major Ridley also reported that he had reached Palapye that morning. The Mounted Infantry were ordered to start for Tuli on the morrow and to arrive there on 26 July. The A Squadron was similarly to proceed to the same place, accompanied by the hospital wagon and Surgeon-Captain Hickson. At Tuli the column was to await the arrival of the Maxim guns and the ambulance.

25 July —Lieutenant Poore's detachment of D Squadron left for Bulawayo at 4 p.m., the remainder of D Squadron under Captain Agnew having arrived at Macloutsie an hour previously. Captain Agnew resumed his march for Bulawayo at 5 p.m. next day, leaving an ox-wagon behind to follow next day with H.S.H. Prince Alexander of Teck.

On 27 July two parties of the 2nd York and Lancaster Mounted Infantry arrived at Macloutsie with the ambulance. There had been no casualties on the road. Two days afterwards Lieutenant Dalgety reached Macloutsie with two Maxim guns and sorry forth thousand rounds of ammunition for them.

A party of York and Lancaster Mounted Infantry left for Bulawayo accompanied by Major Ridley of the 7th Hussars.

At 10 p.m. that night headquarters left for Tuli with two Maxims, an ammunition wagon, a headquarters wagon, and the ambulance. Much difficulty was experienced in crossing the Macloutsie drift as the pull out on the opposite side was very steep and the road generally very bad.

On 30 July they outspanned about seven miles from Macloutsie, where water was good and grazing plentiful. A second stage was made between 3 p.m. and dark and again after 11 p.m., when they marched throughout the night along a road which had somewhat improved though it could by no means be called good. At 10 a.m. 31 July they outspanned at Sinialari spruit, after a march of between seventeen and eighteen miles: again both the water and grazing were good. The march was resumed at 3 p.m. and continued till dark. At 11 they started again and marched through the night; headquarters arrived at Tuli about 9 a.m. on 1 August. There was difficulty at the Shashi drift as the going was extremely heavy. Oxen were, however, hired to take the wagons across while the men remained near the water on the Macloutsie side.

A convoy of six wagons arrived the same day; they were loaded with grain and meal, but had been unloaded at the Transvaal border in expectation of wagons from Macloutsie and Pietersburg which did not arrive. Headquarters left Tuli to join the column that evening, the wagons and Maxims having been sent ahead two hours previously. To complete the supply arrangements Lieutenant Vaughan was left behind. At Tuli there were eighty horses belonging to the Chartered Company, and these they were asked by the Administrator to take on to Victoria.

On 2 August the headquarters rejoined the column at Impaji river, about fifteen miles from Tuli, at 8.30 a.m. A further request from the Administrator was now received by wire asking that all horses and mules at Tuli should be taken on to Victoria. Arrangements to do so were accordingly made and a runner was sent back to Lieutenant Vaughan with instructions. With the exception of one very steep place, about two miles out of Tuli, the road to Impaji was good.

On 3 August they halted all the morning awaiting the arrival of the horses from Tuli; meanwhile Gooding rode on to Umsimbetsi to look for water. A message by runner having been received about 1 p.m. from Lieutenant Vaughan which indicated certain arrangements, the Colonel sent Lieutenant and Adjutant Norton back to meet that officer. Lieutenant Norton started at 2 p.m. and met Lieutenant Vaughan's party two hours later at about one mile from Tuli. The party consisted of two hired ox-wagons, one mule-wagon belonging to the convoy, and also eighty horses of the British South Africa Company. Lieutenant Norton then rode on to Tuli, where he sent wires, and then returned to Impaji, at which place he arrived at 7.30 p.m.

The column had meanwhile started from Impaji at 2 p.m., the hussars finding the advance guard and the Mounted Infantry the rear guard. After proceeding for eight miles the column outspanned at Springs. The march was resumed at 3 a.m. on Tuesday, 4 August, and the road being good, reached the Umsingwane river at 9 a.m. Here Lieutenant Norton rejoined the column. The crossing of the drift was very heavy, as the water was close to the surface of the sand. The Mounted Infantry started getting the wagons across at 2 p.m. After four hours' work three wagons still remained in the drift, and it being dark it was necessary to suspend operations. Moreover, for half a mile on the other side of the drift the road was very bad and exceedingly dangerous. Here news was received that no water was to be found either at the Bullocks Head or in the Umsimbetsi river, and that in consequence Gooding had gone on for another twelve miles to the Umslane river. At 6.30 a.m. On the morrow the crossing of the Drift was resumed. Lieutenant Vaughan arrived with seventy-nine of the horses and the wagons

At 2 p.m. the Mounted Infantry marched, followed an hour later by the hussars. Five native drivers of the Mounted Infantry deserted at this juncture. The march was a very slow one, not owing to the road, which chanced to be not bad, but from the fact that the drivers of the wagon were very unskilled. At 7 p.m., it being dark, the column outspanned, having accomplished a distance of not more than six miles. And now symptoms of rinderpest appeared among the slaughter-cattle—three were in consequence abandoned as they could not keep up. On the road the telegraph party from Victoria met the column and reported that all the natives in that district were quiet.

At 4 a.m. on 6 August the hussars and the convoy resumed their march. The road being very good, eleven miles were covered before they outspanned at the Umsimbetsi river at 8.30. Here, by digging at a spot about two miles up the course of the river, it was found possible to obtain water.

The march was resumed at 4.15 p.m., the distance covered being five miles, but no water was found here. The Mounted Infantry on that day marched half an hour later than the hussars and arrived at Bullock's Head. By digging in the spruit about one and a half miles from the road on the east side, water was found. They resumed their march in the evening and outspanned at Umsimbetsi river. All the oxen but one were now shot, as they had rinderpest.

The march of the entire force was resumed at 5 a.m. on the 7th, and Umslane drift was reached in three hours, a distance of about six miles. Water was found at a spot a mile down stream and south of the road. Here the horses were watered but not the mules. The wagons at once proceeded to Setoutsi, where water was found in the spruit about three hundred yards east of the road. It was, however, rather muddy, and it was difficult to water the animals properly. In many instances the mules utterly refused to drink. At Umslane the column met Heritage, who had not obtained any grain. From the natives with him it was learned that their kraals had been looted by a few wandering Matabele, but that there was no impi in the neighbourhood. The looting was probably the act of some fugitives from Belingwe. The Mounted Infantry reached Setoutsi at 5.30 p.m.

The column resumed its march at 5.30 a.m. on the morrow and proceeded to the first Bubye river. Here the hussars crossed the drift and outspanned beyond the old Post House on the left of the road. The mules of the Mounted Infantry, however, were too tired to go over the river, and were outspanned south of the drift. The distance covered was about six miles. Plenty of good water was found in the pools of Bubye.

The hussars and convoy resumed their march at 2 p.m. on 8 August. They crossed the second Bubye drift, where the sand was heavy, the banks steep on both sides, and the road by no means good. Continuing their march they next crossed the third Bubye drift, and *laagered* up at 6.30 p.m., having accomplished about four and a half miles. One wagon broke down and its load was distributed among the others.

At the *laager* a wagon belonging to the telegraph party and some stores all off-loaded. The Mounted Infantry remained at the first Bubye drift in laager. About 5 p.m. a kaffir was seen signalling from a *kopje*. A party sent out found Mashowa kraal. Some Makalakas who were brought in stated that they knew of no Matabele in the neighbourhood. On 9 August, being Sunday, the column halted. Church parade was held at 9.30 a.m. Later a span of mules was sent back to bring in the wagon abandoned on the previous day. The Mounted Infantry arrived at 10 a.m. At 4 p.m. the hussars and the convoy resumed their march, and after proceeding about six miles halted at 6.30. The road was very good, better than any yet traversed, but the weather was very hot and close during the night, with a slight shower after midnight.

Monday, 10 August.—The march was resumed at 4 a.m. Umsarwe *spruit* was reached, distant about twelve miles. Here there was plenty of water, but the road was not so good as on the previous day. For the last half of the journey it lay between big granite *kopjes*.

Tuesday, 11 August.—A start was made at 5 a.m.; for a mile beyond Kandokwa the road was indifferent but then improved. After six hours' march the convoy outspanned on the south side of the Nuanetsi river, having accomplished thirteen and a half miles. Here the drift was very steep on both sides, with a rocky bottom. The water in the river was running and there was plenty of it. The wagons occupied two hours in crossing, and halted for the night on the river bank.

Here an accident befell Armourer-Sergeant Baldock; an empty hogshead fell on his left hand from a wagon, cutting off the little finger and severely injuring the third finger.

August 12.—The march was resumed at 5.30 a.m. to Makalakask kop. About three miles on a spruit was crossed, where there was plenty of water. Six miles farther on a large pool was discovered about one mile east of the road behind a kopje. Evidently there was plenty of water about, but it was not easy to locate it. The mules were watered at the pool, but for some reason the horses were sent back to the spruit. The march was resumed at 3.45 p.m. and continued till 7 p.m., the distance covered being about seven and half miles and the road good. On the way the *bush* of a wagon was found to be worn out, but it was cleverly patched by Private Winsborough by utilising the spokes of an old wheel.

13 August—The hussars and the convoy resumed their march at 6 a.m. to Lundi river, a distance of about fourteen miles. The road was good, but there were three bad drifts to cross. At Sugarloaf, about seven miles from Makalaka kop on the left of the road, there was plenty of water. At Lundi river there was a strong stream of good water, but the drift was very heavy going. At daybreak Gooding, who had remained at Makalakas kop, went back to meet the Mounted Infantry and to show them the water. He returned to the column with news that they were getting on well and hoped to reach Lundi that night. The wagons crossed the drift at 4 p.m. The Colonel and Lieutenant Norton now went ahead to Victoria, taking with them three servants and starting on their journey about 8 p.m.

At Lundi, Trooper Usher had been found awaiting the column with a supply of grain which had been sent out by the Civil Commissioner at Victoria. The Colonel and Lieutenant Norton reached Pollock's Store on the Iokwe river about 10 a.m. on the 14th and rested there till 6.30 p.m.; their march was then resumed till 10 p.m., when they off-saddled for the night a short distance beyond Fern spruit. On the morrow they reached Fort Victoria about 10.30 a.m. Here the entire population went into *laager* nightly at 9 p.m., the fort itself being manned by about one hundred burghers and volunteers, who had also a Gatling gun. From any native attack the fort was found to be quite impregnable. The Civil Commissioner stated that all the natives on the Charter road beyond forty miles from Victoria were rebels, but that it was difficult to ascertain whether they had actually committed any murder or outrage. It appeared, however, that the Native Commissioner Drew had been attacked on this road a few days previously and had with difficulty escaped from death.

On 17 August the Colonel interviewed all the Native Commissioners and wired a digest of the scanty information obtained to the C.S.O., Bulawayo. This may be summarised as follows: Chaka, Gambisa, and Umtigeza were unfriendly. M'Takinompie, the Chief of the Hartley Hill District, was by report responsible for most of the murders round Salisbury, and had fled to Taba Enzembi, near

Charter. Charter was stated to badly need provisions, and instructions were asked for as to revictualling it; the Colonel also requested general instructions as a guide in his dealings with unfriendly chiefs. The murder of a white trader by Umtigeza's people was suspected, but direct evidence of the deed was lacking. On Tuesday, 18 August, A Squadron arrived about 8.30 a.m., having lost one horse from horse-sickness.

In answer to the Colonel's wire the following telegram was received from Sir F. Carrington:

> I think it undesirable to attack any chief who may at present keep quiet, although at heart unfriendly, as it extends the theatre of operations, and they can always be dealt with later when the heart of the rebellion is well broken. I intended that when the present force in Mashonaland could clear the upper part about Salisbury, Mazoe, and Lomagundi, they should afterwards turn their attention to Charter, Mashigombi and Hartley Hill. If there was no pressing work for you to do near Victoria or in the Victoria District, I wanted you to march to Gwelo, as that country has not been operated in, only a small garrison there holding the place. As soon as Ridley returns from his present patrol I intend sending him and the hussars, and 56 Mounted Infantry and 150 Police, across the Shangani to operate against the rebels who are in force between Shangani and Gwelo, and this would be about the time you would be operating much in the same district from Gwelo. The Proclamation concerning the surrender of rebels is not yet applicable to Mashonaland.

On Wednesday 19 August, the Mounted Infantry marched in about 9 a.m. News arrived from Bulawayo that Charter had been relieved from Salisbury and that therefore there was no urgent need for any action on the part of the column. On that day all the wagons were sent to the blacksmith's to be generally overhauled. Next day news arrived that the rebels were on the Kwekwe river near Gwelo, and instructions were received that they were to be cleared out of that locality. The information that Charter and Elkedoon had been relieved was confirmed.

On Friday, 21 August, a message by wire was received from Gwelo to the effect that native scouts who had been sent out thence three weeks previously had returned and reported that rebels with cattle were in a number of small kraals on the Kwekwe river, which crosses the Gwelo and Ironmine road and runs north. The report was confirmed from Fort Gibbs, as the spoor of between six and seven hundred rebels had been seen crossing the road in the direction of these *kraals*. It was also reported that rebels had taken refuge on the lower Gwelo river and were probably under Manondwan and Kwazikwazi. The trouble now was to decide which was the best road to take to Gwelo. Of roads there were two, and the accounts of them were very conflicting. That *via* Selukwe was the shorter, but was reported to be very bad in places, and it was also stated that the grass had been burnt on both sides of the road. The other, *via* Ironmine Hill, was stated to be well watered.

That evening the column was given an entertainment by the people of Victoria. Since their arrival at Victoria there had been trouble with the horses and mules, which for no apparent reason were off their feed. Here Lieutenant Holford had the misfortune to lose his horse John Davis from horse-sickness. Next day rations were issued to complete the column to forty-five days' supply from the 19th. A redistribution of wagons was made also. At 6.30 that evening a heavy thunderstorm with rain, which continued throughout the night, broke over the camp, but as fortunately tents had been pitched on the arrival of each party, not much damage was done.

On Sunday, 23 August, at 7 a.m. seven wagons were loaded. The rain still continuing, church parade was countermanded. During the afternoon the remainder of the wagons were loaded, except two belonging to the Mounted Infantry, which were not yet repaired.

On Monday 23 August, the column left for Gwelo at 4 p.m., marching by the Selukwe road, twelve volunteers and fifty natives with a few police, under Lieutenant Forrestall, accompanying them. The rate of progress was very slow. At 6 p.m. a wagon turned over in Victoria spruit about four miles out. The force then outspanned for the night, during which there was some heavy rain. Next day the march was resumed at 6 a.m. The wagons did not all get over the spruit till 11 o'clock. The column marched on till 2 p.m., but only covered three miles as the rain had made the roads very heavy indeed. At 8 p.m. they started again and went on till midnight, when a rough *laager* was formed.

Wednesday, 26 August.—Marched at 6 a.m. Reached Loot *kraal* about 11 a.m. and were now 15 miles on their way. The mules, however, were all knocked up owing to the heavy roads. The drift, which was a bad one, was improved by dynamite before crossing, after which the force *laagered* on the other side. At 5 p.m. there was a parade under arms for manning the wagons, and this duty was a daily order for the future. The march was resumed at 9 p.m.

The column outspanned at 3 a.m. on 27 August. The Mounted Infantry wagons were very late. One was left behind with a broken disselboom and did not get in till 6 a.m. The hussars resumed their march at 7.30 and reached Shashi at 9.30 a.m. The drift was rather bad, but the wagons managed to cross. At 4.30 p.m. the Mounted Infantry arrived, and the *laager* was completed at 5.30 p.m. An alarm was sounded at 7 p.m., and the wagons were quickly manned.

On Friday, 28 August. the march was resumed at 4 a.m. Lieutenant Norton rode ahead with Gooding (starting an hour earlier) to improve the road. The drift over Umjezi was found to be very good. Three miles farther on water was found. The column arrived about 10 a.m., having achieved five miles. At 2 p.m. Southey's wagons were sent ahead. Help's followed them half an hour later, and the remainder marched at 3.30 p.m. The force outspanned at about two miles from Tokwye at 6 p.m. A section of Mounted Infantry was now sent on to escort some grain wagons, and theses were outspanned about one and a half miles on. Lieutenant Vaughan, with his troop and the Victoria contingent and natives, were now despatched ahead to raid Chinus *kraal* across the Tokwye.

Saturday 29 August.—The column marched to Tokwye at 3 a.m., crossing a very bad drift. They *laagered* on the far side. Lieutenant Vaughan returned at 10 a.m. and reported that all the natives had been cleared. The march was resumed on Sunday. 30 August, at 3 a.m. The road was good and by 9.30 a.m., having accomplished nine miles, the column halted and *laagered*. Second Lieutenant Greville took a troop on patrol to examine a *kraal* on a hill called Spitzkop. This *kraal* had been burnt by the Indaima patrol, but the natives had returned to the spot and were rebuilding it. They all cleared on sighting the column. The *kraal* was again burnt. About 11 a.m. some natives came in and reported that a man had shouted to them from a cave. Lieutenant Norton went up to the cave with Forrestall and the native police and fired some dynamite shells inside but with no result. The march was resumed at 2 p.m., and having proceeded about three miles the column *laagered* at 5.15.

The next day (31 August) was more exciting—the march was resumed at 3 a.m. Just after daybreak some *kraals* were seen on the right of the road. Second Lieutenant Holford was accordingly sent up to them with his troop. He found the place deserted, the *kraals* having been burnt, but there was a considerable supply of grain there. The column outspanned on the far side of the Little Umtibekwe river at 9.30 a.m. Here one section of Mounted Infantry was left in laager. The remainder of the column then went out on patrol to endeavour to intercept the natives who were clearing from Makamisi's *kraals*. The country was found, however, to be impossible for mounted men, and although a few natives were seen nothing could be done. A section of Mounted Infantry was sent out under Lieutenant Tyler against a stronghold on a hill to the west of the road at 3.30 p.m. They sent back word that the force was not strong enough to carry the position, so the Colonel took up twenty-five hussars and twenty-five Mounted Infantry. There was a little firing, but all the natives had cleared or taken to caves, having first set fire to their kraals. The force then returned to *laager*, which was reached at 5.30 p.m.

On 1 September at 3 a.m. the column marched, crossed the Umtibekwe river, and laagered at the spruit about three and a half miles on. Under the escort of a troop of hussars the grain wagons were sent on to Selukwe. *Kraals* on a hill near the camp were burnt by the natives, one man was shot, and some meal and eggs were obtained. A colonial boy, who had been for eight months with the Kaffirs, came in and surrendered. He stated that the rebels were collected near the Dunraven Mine. Ten police under Sergeant Walters also arrived from Gwelo.

At 3 a.m. on Wednesday, 2 September, the wagons started. A column of fifty hussars and fifty Mounted Infantry were sent out to Sinkwas *kraal* to look up the natives the colonial boy had mentioned. The party reached a hill overlooking the *kraal* about 6 a.m., but found it had been deserted for at least a week. No grain or loot of any kind was to be obtained. The party consequently returned to the wagons and outspanned. All the stores along the road had been looted by Kaffirs, and a great quantity of dynamite and numbers of candles were lying about. The night was very wet, but the column started at 4 a.m. and marched for nine miles. Second Lieutenant Holford's troop, with the contingents from Victoria and Gwelo and the natives, visited the *kraals* of Monogola about four miles from Bonsor Mine. A few natives were seen and there was a little firing. One of Holford's natives was shot through the body from a cave while climbing at the top of the *kopje*. The cave was at once blown up with dynamite, but the *kopje* being full of caves not much could be done. Grain was found in bins in the *kraal*. While returning to the *laager*, as the men passed a *kraal* near, which seemed to be deserted, Trooper Birand of the Victoria contingent was shot in the left shoulder, the bullet passing into his lung. A native fired this shot from a hut, and then hid in a cave whence it was impossible to dislodge him. The wounded men were carried back to camp and reached it at 4 p.m.

On Friday, 4 September, the column arrived at Gwelo at 10 a.m. The last march was fourteen miles and the start took place at 6 a.m. The road was found very good, though the night had been rather wet. The wounded native died on the night of 3 September. The officer commanding at Gwelo was Captain Pocock. His command was taken over on 5 September by Major Thorold, who arrived for that purpose, and with Major Thorold came nine wagons of foodstuffs.

On Sunday, 6 September, a church parade was held at 9.30 a.m., and White's scouts left for Bulawayo. Next day the wounded Trooper Birand died and was buried with military honours, the hussars finding the firing party. Preparations were made for a projected patrol which was to start at 5 a.m. On 8 September.

Tuesday. 8 September.—The patrol in the Kwekwe District now took place. The column started at 6 a.m. Grain was, however, very scarce and the allowance to the hussars was reduced to 6 lb. *per diem*— that to the Mounted Infantry being 5 lb., while the mules had no more than 2 lb. The first march covered a distance of about nine miles, and the column then laagered. The road was rough but on the whole not a bad one. Scouts went out in al directions, but found that all *kraals* had been for sometime abandoned.

Next morning the column started at 5.30 a.m. And proceeded for seven miles. The road was good, and the latter part lay between castellated *kopjes*. The spoor of several small parties of rebels was found but no large bodies had apparently been that way. A patrol followed the spoor and returned after about an hour with one man, one woman and two children prisoners. Another patrol sent out in a different direction captured a baggage bull, but the men with it escaped. From the captured man it was learnt that only small parties were about, and that these were seeking for grain. The march was continued from 2 p.m., and proceeded for about six miles towards the Kwekwe river, when it *laagered* on the bank of a spruit. Private McGeorge of the 7th Hussars, having been guilty of disobedience of orders in lighting fires in the long grass—a most dangerous practice and strictly forbidden—was tried by a field general court-martial and sentenced to fourteen days' field imprisonment No. 2.

Thursday, 10 September. The column started at 5 a.m. and turned off Hunters' Road on to the Phoenix Mine road, marching to Kwekwe drift. A *laager* was formed about half a mile on the near side, the distance covered being nine miles. Patrols of fifty hussars and fifty Mounted Infantry preceded the column on the right and left of the road respectively, and reconnoitred the country across the Kwekwe. The hussars came on a *scherm* about 5 a.m. and captured nine women and several children; the other party sighted nothing. Some Martini and Metford ammunition was found by the hussars. Men had evidently been there but had made themselves scarce, having escaped into the bush. The women reported that most of the natives had moved to the Surnamboola forest. At sunset Lieutenant Vaughan went out with a patrol, returning at 10 p.m., but beyond sighting several fires nothing was discovered.

On Friday, 11 September, the march was resumed at 5 a.m., but merely to return to the site of the outspan of Wednesday. Patrols went out as before at about 4.30 a.m. The hussars went back more or less over the same country and found a child that had been left behind by the natives. The Mounted Infantry moved for some distance up the Kwekwe river, and then crossed to Hunters' Road, and thus back to camp. They saw nothing. The force remained in laager. This day the Commanding Officer released McGeorge.

Saturday, 12 September.—The column remained in *laager* during the morning, but fifty Mounted Infantry were sent on patrol to some *kraals* about ten miles away. They returned at 1.30 p.m., having found no grain, though there were traces that a considerable quantity had been quite recently removed. At 4 p.m. the whole column started, and having marched about five miles outspanned at 6 p.m.

Sunday, 13 September.—The column started at 5 a.m. and marched to Harboard's store, distant some six miles. Patrols were out on both sides of the road and found *kraals* deserted, but twenty sacks of Kaffir corn and thirty-three sheep and goats were discovered. Near the store the remains of a man named Fitzpatrick and those of another, supposed to be Hartley, were found. They were buried in one grave. A patrol of one hundred and fifty men now started for Matoro *kraals*, about fifteen miles away. Here several chiefs were reported to be in the hills, their names being Manandwan, Guibana, and Quazi Quazi. The patrol halted for the night at a spot about four miles from the *laager*.

Monday, 14 September.—The column marched at 5 a.m. All *kraals* were found to be deserted, the natives having cleared on the approach of the patrol. Some grain was found in one *kraal*, and two wagons were sent for to carry it in. At 11 p.m., the horses having been grazed patrol marched to Gwelo river and halted for the night. Several women were captured and one man was shot by the scouts.

Tuesday, 15 September.—The force started at 5 a.m., scouts being sent on to reconnoitre the Matoro hills. The column halted at the corner of the hills where some *kraals* had been burnt on the previous day. When the scouts returned they reported that some *kraals* containing grain had been discovered. At 8.30 a.m. one wagon, sent for on the 14th, arrived. Half the force was now sent on to the Gwelo river, distant about three miles, to water, while the remainder took the wagons on to the *kraal* and loaded them up. The watering party then joined them at the *kraal*, and the other half were sent to water, with orders to return by 5 p.m. The Colonel now decided to return to the *laager* and not to attack the hills on that day. At 4.45 p.m. The videttes gave the alarm, and a number of natives were seen advancing towards the *laager*, but they turned out to be only a party coming for water. The scouts killed one man. Half an hour later the force started back to the *laager*, where it arrived at 10 p.m. As soon as the troops were clear of the hills the captive women were released.

Wednesday, 16 September.—The column remained in *laager* till 5 p.m., and then started for Gwelo drift. After a march of seven miles they *laagered* for the night. A patrol from Ridley's column, consisting of Captain Kekewich (2nd York and Lancaster Regiment) and de Moleyns (4th Hussars), with twenty-seven non-commissioned officers and men, arrived at the *laager* at noon. They had found that the natives were everywhere dispersed in small bodies. It was arranged that they should accompany the column to the Matoro hills. There was a report that the chief Manandwan had cleared into the Sommabula forest.

Thursday, 17 September.—The column started at 5 a.m. and marched about two miles to the Gwelo drift. Having crossed the drift, they *laagered* on the far side, good water being found in the pools. A patrol of one hundred and fifty men and the York and Lancaster Mounted Infantry, who joined yesterday, taking with them one Maxim gun, started with three days' rations at 6 p.m. Halted for the night at 10 p.m. On the hills various fires were visible.

Friday, 18 September.—The column started for the hills at 3 a.m., and on arrival at the end of the range at daybreak it was found that all the natives had cleared. One man was killed by Kekewich's men while attempting to escape. At the foot of the hills the axle of the Maxim broke, and the column then returned to the drift, where it waited till 3 p.m., when it returned to the *laager*, a distance of about five miles. One section of mounted infantry under Watson remained behind to accompany Kekewich's patrol in an expedition against some *kraals* near Lion Kopje, a spot about twenty miles down the Gwelo river. This force was to follow the column afterwards to Gwelo. On this day Captain Rivett-Carnac (2nd West Riding Regiment), who was suffering from dysentery and eczema, was placed on the sick list. Two wagons were sent on to follow Watson's patrol.

Saturday, 19 September.—The column started on its return march to Gwelo at 5 a.m. A prisoner having informed the Colonel that grain was hidden in pits at a spot about ten miles south of the road, Lieutenants Vaughan and Forrestall, some natives, and twenty men started at 4 a.m. to look for it. The column *laagered* after a four hours' march. A messenger arrived from Lieutenant Vaughan about 11 a.m. with a report that he had not yet discovered the grain, but had killed three natives. At 2.30 p.m. another report arrived.

The original of the second written message (in pencil and very faded) being in existence, it will be interesting to insert it here. Unfortunately it was not sufficiently clear to reproduce in facsimile. It is as follows:

Posn. 4 miles S. Gwelo river. 10 from last night's halt. 12.30 a.m. The two natives who take this in will show the wagons the way to the grain. There are about 10 bags, but send two wagons as they may find more, send water . . .(illegible) and small bags to get the grain. There is no water within 4 miles. The other two natives will bring me an answer here. After leaving the first scherm (?) as reported I killed 15 natives. I don't think any men escaped. The spoor of most of them crossed more to the west with a few oxen and one mule. I had to leave this to get to the grain. If the C.O. permits I will follow them up tonight. I have 8 men rationed till tomorrow night, 12 till tonight. I should like a few more biscuits and coffee sent out, otherwise can manage as the natives have meat. The men killed were coming towards me from the south have nothing to do with those with the oxen. If you send any more rations, send it to the water by natives, the whole gang can then go up to the grain place. I enclose a rough sketch. When I return I will follow up your wagon spoor. If the C.O. wishes me to come back I will start about dusk. If I may go after these niggers I shall start about 9 p.m. and hope to catch them in the morning and will follow your spoor back towards Gwelo catching you up as soon as possible. Please say there is nothing to fear from these people, as those who have arms throw them away when spotted. Please don't send any more men as the water holes are only sufficient for my party.
J. Vaughan

At 3.30 p.m. two wagons with one day's rations and an escort of ten men were sent out. As several mules were missing the column did not march this evening. On this day there was no grain for the animals.

On Sunday, 20 September, at 5 a.m. the column marched for Harboard's old store, arriving at 9 o'clock. Again throughout the day the animals were without grain. A patrol of eighty men was then prepared to start for the Tumain hills, where it was reported that Guibana and Quazi Quazi were to be found in hiding. The patrol

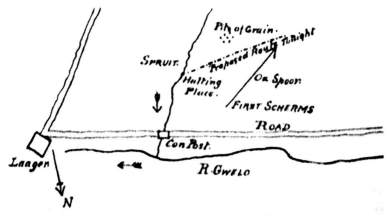

started at 2 a.m. on the morrow, and reached the kraals at daybreak. All, however, were found to have been deserted for some time. They proceeded for some distance to some water holes, but these were dry, and the patrol returned to water and off-saddle. At 11 a.m. some women were captured who had come in quest of water. From them it was learned that Guibana had cleared with some cattle two days previously, and that he had taken refuge in the Senonombi; they also stated that they could point out some grain in pits. Taking one of them as a guide a patrol went out but found no grain. Near the water at Smith's camp three bodies were found. Near one of them was the diary of Albert T. Lee, sailmaker, Bulawayo. The second skeleton was probably that of a Cape boy; while the skull of the third was broken into fragments. The patrol started for the *laager* at 6 p.m., and on arrival found that Lieutenant Vaughan had returned during their absence and brought with him ten bags of grain.

On Tuesday, 22 September, the column remained in laager all the morning. The wagons started at 4 p.m. escorted by a troop of hussars for Marin; the remainder marched at 5 p.m. independently.

Next morning at 4 a.m. the column marched to Five Mile *spruit*, where it outspanned for the day. Starting again at 4 p.m. Gwelo was reached at 7.15 p.m., and they *laagered* about one mile from the fort. At noon Watson's patrol met them, but had only been able to obtain grain enough for themselves; they had, however, accounted for twenty natives.

At Gwelo on 24 September news arrived that 'Ndema, so far from being crushed, had refused the terms offered to him by the native commissioners. On the next day, in response to a reference to the General Officer Commanding, it was ordered that 'Ndema should be attacked without delay. The condition of the horses was at this time far from satisfactory, owing to the lack of grain; they were in fact nearly starved.

On 26 September the horses were sent out to graze about two and a half miles on the Selukwe road.

The column rested at Gwelo until 2 October. Two days previously sixty-six horses and Mounted Infantry ponies were found quite unfit to proceed on the projected patrol, and were therefore sent out to Brickfields Camp, where the grazing was better.

On 1 October rations up to the 16th inclusive were drawn for the column, and five wagons with spans were handed over to the officer commanding at Gwelo.

On Friday, 2 October, the patrol against 'Ndema marched at 4.30 a.m. and halted for the day at Half Way House. Here a report of the movements of some natives was received from Mr. Gilfillan, who occupied a farm at Wareleigh, and in consequence a patrol consisting of one troop of hussars and one section of mounted infantry started at 7 p.m. for Dunraven Mine. The patrol halted for the night near the old police camp, which was not far from Allen and Maclean's store. Colonel Paget, with Mr. Gooding, went out beyond Dunraven Mine, but could see no signs of natives. Next day the patrol marched at 5 a.m. to Tinkwas *kraal*, which had been destroyed by the column a month previously, but a wrong direction was taken, and the horses being bogged and unable to cross some of the valleys, which are very steep with boggy bottoms, the patrol rejoined the wagons at Weir's Store at 1 p.m.

Sunday, 4 October.—At 2 a.m. a patrol of ten mounted and twenty dismounted hussars with two wagons, under Captain Carew, left camp, their destination being Monogola's *kraal* near the Bonsor Mine. The patrol surprised the stronghold, killing three rebels, one of whom was a Cape boy. They were unable, however, to take the caves, and being under fire from the neighbouring *kopjes* without being either able to see or to get at their enemy, the patrol withdrew. In rushing the *kraal* Lance-Corporal W. A. Smith 7th Hussars) was killed, being shot through the head at very close range out of a cave. One of the native police also was slightly wounded in the arm. Meanwhile the remainder of the column had attacked a number of *kraals* at the south end of the Selukwe hills. About twelve *kraals* were destroyed, the enemy deserting them without resistance as the troops approached. Only one shot, indeed, was fired by them. These *kraals* belonged to Msengwa. About twenty-five sacks of grain were obtained and were very welcome. The wagons had moved on to the Umtibekwe drift. The Mongolas patrol returned at 4.30 p.m. bringing with them the body of Lance-Corporal Smith.

The funeral of the unfortunate man took place on the morrow. Patrols went out to endeavour, with the assistance of the natives under Native Commissioner Forrestall, to obtain more grain from the hills visited yesterday, but were not successful.

On Tuesday, 6 October, Colonel Paget, with the officers commanding the hussars and the Mounted Infantry, left camp at 3.30 a.m. to reconnoitre Magomisa's stronghold, and meanwhile three patrols were sent out as follows: No. 1, ten men under Lieutenant Smith to meet the native contingent from Victoria. No. 2, a hussars patrol under Lieutenant Vaughan, which went south of the road and captured the Bugai hill, being fired on but without sustaining any casualties. Squadron Sergeant-Major Willard and Farrier-Sergeant Grey got lost, but after a night's wandering succeeded in finding their way back to camp on the morning of the following day. No. 3, patrol of mounted infantry, was under the command of Lieutenant Tyler and operated on the north side of the road. This patrol captured ten sacks of grain. A wagon was sent back to Gwelo under escort of one sergeant and ten men, and carried three sick men.

Early next morning a search party was sent out to find the two lost men, but did not succeed in so doing, as the two returned to camp during their absence. A patrol was also despatched to bring in the ten sacks of grain found by Lieutenant Tyler on the previous day. As there was better grazing ground higher up the river three miles off to the north, the wagons were moved thither. On the previous day a message had been brought by runner from Gwelo. This message had been sent thither by Lieut.-Colonel Baden-Powell from Shangani at 8 a.m. on 5 October. It ran as follows:

From Baden-Powell to Paget
'Ndema's
3rd October
Am moving with 100 mounted men to co-operate with you against Wedza, where could arrive 7th or 8th. am anxious to have instructions as soon as possible by runner to Belingwe road along which I shall be moving. I have one 7-pounder and 2 Maxims. Am rationed to the 22nd except grain, of which I have only a few days' supply. If you have not yet finished with 'Ndema I might possibly help you there. I reach Inseya Drift to-morrow.

To this message a reply was sent by runner to Gwelo to be wired to Shangani and sent thence by runner to Belingwe road

7th *October*
Lesser Umtibekwe Drift
Please cancel my message despatched last night if you receive it before this.

This was a message to say that the force would go to meet Colonel Baden-Powell but it was not sent, being cancelled by the request of the senders on arrival at Gwelo, a mounted messenger having been despatched to intercept it. The message of 7 October continued to the effect that the work with 'Ndema could not be quitted in order to go southward to Belingwe road without positive orders from the General Officer Commanding. The column would therefore be working northwards towards Iron Mine hill, and it was suggested that Colonel Baden-Powell could best co-operate by clearing the country between Belingwe and Lundi.

Colonel Baden-Powell was asked to send his 7-pounder to the use of the column at the conclusion of his patrol, it being urgently required to deal with Monogola's stronghold, which had to be taken as soon as 'Ndema was finished with. There were only rations sufficient to last till the 16th, and it was particularly desirable that the operations then in hand should conclude before the rains began, as after that, mule transport would be impossible in many places owing to the soft soil. A duplicate message was also sent to the General Officer Commanding at Bulawayo.

On Thursday, 8 October, all the mounted men and a Maxim proceeded to Magomisa's stronghold, starting at 3 a.m. This worthy was stated to be an adviser of 'Ndema, and it was therefore important to attack him at once. His stronghold was a difficult place to deal with, having cultivated lands and a spring of water on the top of the hill On arrival of the column, however, it was found to be deserted. A few goats and two head of cattle only were captured there.

During the afternoon the Native Commissioners Weale and Eksteen arrived from Victoria with two hundred of Chilimanzi's men, and Magomisa's *kraal* and four or five others on the same Selundi range were destroyed by them, but only a small quantity of grain was found. A child who had been left behind in Magomisa's *kraal* stated that the rebels had gone to the Selukwi hills. A runner arrived on the same day from Gwelo, bringing a copy of orders dated 4 October for the co-operation by Colonel Baden-Powell with the column against Wedza as soon as 'Ndema's patrol was concluded. In consequence a wire was sent to Baden-Powell at Belingwe road giving the substance of the message, and adding that it was not known when 'Ndema would be finished with, but that the column would possibly reach Wedza about 20 October, and that they would bring with them seven days' rations for Baden-Powell if such could be obtained.

The column intended to approach from the direction of Shashi drift on the Gwelo Victoria road, and had settled to send a patrol ahead to communicate with Baden-Powell.

On Friday, 9 October, the column marched towards 'Ndema's stronghold of Sika, starting at 5 a.m. Patrols were sent out on both sides of the road. About six miles west of Sika a halt was made and a *laager*. Up to this point the track across the *veldt* was quite easy for wagons in dry weather, but after heavy rain would have been quite impracticable.

The patrol on the right under Lieutenant Tyler found some grain, and two wagons were sent with an escort to bring it in. They returned on the morrow with forty sacks, mostly of small grain. Owing to the guide losing his way in the bush these wagons were delayed. A wire was sent to Colonel Baden-Powell on this day making arrangements for the meeting, and asking him to come out to meet the advancing column, as it might be delayed by accidents connected with the road, which was not well known but which was believed to go from near Shashi drift on the Victoria Gwelo road to Belingwe. It also informed him that a rocket would be sent up nightly within an hour after sunset after crossing the Lundi until he was either met or heard from.

Saturday, 10 October.—The wagons remained where they were under guard. The remainder of the force marched at 4.30 a.m. and reached 'Ndema's stronghold of Sika at 6 a.m. A few rebels were seen on the hill but they made off, offering no resistance. The abandoned *kraals* were destroyed and thirteen head of cattle were captured by Forrestall's One old man was taken prisoner, who stated that 'Ndema was anxious to come in. Many other *kraals* were destroyed on that day. At 4 p.m. the old prisoner was sent with a message to 'Ndema to come in, but he returned next day to say that 'Ndema was afraid and had gone to Banka's across the Tokwe. He also stated that the natives had two *indabas*, one party being in favour of surrender, and the other of going in a body into Chibi's country In the evening Forrest all returned to Victoria with his police and boys. Private C. Usborn of the 7th Hussars died in Gwelo Hospital of meningitis.

Sunday, 11 October.—The column remained in *laager*, sending out patrols to the north, east, and west. Lieutenant Vaughan obtained twenty-four sacks of grain, and Lieutenant Holford visited Senangwe, formerly 'Ndema's principal stronghold. He found the place deserted.

Monday, 12 October.—The force marched south-west at 4 a.m. to 'Ndema's old *kraal*, and resumed the march in the evening to the Victoria Gwelo road, passing over extensive lands; but no grain was discovered. That night a runner arrived from Gwelo bringing a message that Colonel Baden-Powell had been ordered to join the column and assist against 'Ndema. Accordingly a message was sent to Colonel Baden-Powell asking for information as to his movements and the date of his probable arrival in the neighbourhood of Gwelo, and stating that the column would probably operate on the Selukwe District till his reply was received, after which instructions would be sent for combining their movements. The column marched at 5 a.m. on 13 October, patrolling each side of the road till a halting place was found where there was water and grazing; this spot was about three miles in the direction of Victoria. 7000 lb.

of grain was discovered on the right and about 2000 lb. more on the left. A small patrol under Lieutenant Greville was sent out to meet the wagons for Gwelo, which were coming out with rations.

On 14 October the force remained in *laager*, but some patrols sent out obtained more grain from the same district as on the previous day. Next morning the *laager* was moved back to Little Umtibekwe drift. On the march one wagon broke down, a wheel going to pieces. The two wagons from Gwelo with rations arrived. A message from Colonel Baden-Powell announced that he was at Possel's Farm awaiting instructions. The reply asked him to join the column at Wareleigh, and informed him that Captain Watson would be sent to meet him at Lundi with seven days' rations, if he came that way. Captain Watson accordingly started next morning with Mounted Infantry and seven wagons to the Lundi and was instructed to wait there till the 22nd and then to proceed against Banka, with whom 'Ndema was stated to have taken refuge. He was then to rejoin *via* the Iron Mine Hill road. The laager was then moved to Sinkwa's Gardens. A patrol sent up the Umtibekwe valley found a small *kraal* and destroyed it.

Saturday, 17 October.—The column marched to Weir's Store. A mounted patrol of thirty hussars with two days' rations was sent under Captain Carew to operate in the Selukwe hills.

Sunday, 18 October.—The column marched to the Half Way House (Wareleigh), arriving at 8.30 a.m. Three sick men were sent on to Gwelo. The patrol sent out on the 16th returned during the afternoon, having seen nothing. A broken wagon was sent to Gwelo for exchange or repair. Next evening the Colonel and Adjutant with the headquarter wagon proceeded to Gwelo. Private L. Tay of the 7th Hussars died of enteric in Gwelo Hospital.

20 October.—News arrived from Colonel Baden-Powell. He had been operating between Belingwe and Lundi. He reported that they had 'knocked out Wedza yesterday' and that his people had fled towards the column. He adds:

> If you could kick them back again it would have an excellent effect. They are on the hills along the Sabi N. E. of Belingwe (Meikle's Store).

It appears that the natives south of the Victoria-Belingwe Road were friendly. Matzetetza's stronghold and a few other *kraals* had

165

been destroyed. Colonel Baden-Powell was then on the Singweze river at a spot fifteen miles north of Belingwe. His force amounted to seventy men and horses fit for work, but only half the men had boots, and the supply of rations would only last until 4 November. He awaited further orders. In reply Colonel Baden-Powell was informed that the column would remain at Wareleigh. Private Tay was buried in the cemetery at Gwelo at 11.30 that morning.

Wednesday, 21 October.—The Colonel rejoined the column at Wareleigh. Nothing occurred on the 22nd and 23rd. On the 24th, however, Native Commissioner Driver arrived in camp bringing news that 'Ndema had sent in a message to the effect that he wished to surrender, together with Magomisa, Makonisis. and Banka. In reply the chiefs were told to come to Weir's Store on either the 25th or 26th. Native Commissioner Driver accordingly went thither accompanied by Lieutenant Norton, one sergeant, and six men. They stayed the night at Weir's Store. On 25 October they rode out to Dunraven Mine. No signs of Ndema at Weir's Store. They again remained at the Store for the night.

26 October.—Lieutenant Norton rode out to the mine again at 5 a.m. Three hours later the column arrived from Wareleigh and *laagered* to await the arrival of Colonel Baden-Powell's column. At 4.30 p.m. Colonel Paget rode out to meet Baden-Powell and found his force *laagered* at the Tebekwe mine. Colonel Baden-Powell's column arrived at Weir's Store at 7 a.m. on 27 October. Its strength was as follows:

Officers: Lieut.-Colonel Baden-Powell; Major H. M. Ridley, 7th Hussars; Captain C. H. Agnew, 7th Hussars; Second Lieutenant A. Imbert-Terry, 7th Hussars; Second Lieutenant H.S.H. Prince Alexander of Teck, 7th Hussars; Lieutenant H. P. Thurnall, 2nd Yorks and Lancs Regiment; Surgeon Captain Ferguson, A. M. S. N.C.O.s and Men: 7th Hussars 34; 2nd Lancs Regiment 64; M. S. Corps 2. Total 100. Horses: Officers' 8; 7th Hussars 26; 2nd Yorks and Lancs 63. Total 97.

28 October.—The column marched at 4 a.m. by the Bonsor Mine road towards Monogola's *kraal*. Lieut.-Colonel Baden-Powell, Captain Carew, and six men went on at 3 a.m. to reconnoitre. They returned at 8 a.m.. having found the *kraal* apparently deserted. The column *laagered* on the Shangani road about one mile from the Bonsor Mine. At 3 p.m. another patrol went forward with twelve

mounted and fifty dismounted men, the 7-pounder and a Maxim, and thoroughly explored the stronghold. It had evidently been deserted for some time, probably from immediately after our last visit.

29 October.—At 5 a.m. Lieutenant Norton returned to the kraal and explored the caves, which were of great extent. About fifteen sacks of grain were discovered hidden. A party was left to get the grain out and the rest returned to the *laager* at 10 a.m. Colonel Paget with Lieut.-Colonel Baden-Powell, Surgeon-Captain Hickson, and Lieutenant Norton left for Weir's Store, *en route* to Wareleigh, at 4 p.m. The column marched to Monogola's *kraal*.

Friday, 30 October.—Marched to Wareleigh at 5 a.m. The column under Major Ridley remained in *laager* at Monogola's, and sent out patrols to search for grain and to endeavour to find the spoor of natives. Next day Colonel Paget proceeded to Gwelo, where he arrived at 10.30 a.m. Major Ridley, having procured all the grain which was to be found in Monogola's caves, blew up the stronghold with thirty-seven cases of dynamite which he obtained from the Bonsor Mine. He then returned to Gwelo. The dead bodies of nine natives were found at Monogola's stronghold; they had evidently been killed by the last patrol.

On Sunday, 1 November, a telegram was received from Bulawayo ordering all the 7th Hussars to proceed by easy marches to that place as soon as the patrols were over. Special instructions regarding the mounted infantry of the 2nd West Riding Regiment were to follow. Until these were received they were to remain in or near Gwelo. When their work was finished the present arrangements were for all the 7th Hussars then in the country to remain in Bulawayo during the summer. A permanent camp was being prepared for them. The mounted infantry of the 2nd West Riding Regiment, if not required for other duties, were probably to march down the country shortly.

On 2 November the force remained quietly at Gwelo. Next afternoon a telegram arrived from Bulawayo reporting that a few hundred Matabele and Mashonas and some of Umtigesa's people were said to be occupying Manezain Intaba Zimbi or Iron Mine hill range, about seventy miles from Gwelo. A column of hussars and mounted infantry of a strength of two hundred were to proceed thither at once. Either Colonel Paget or Major Ridley was to

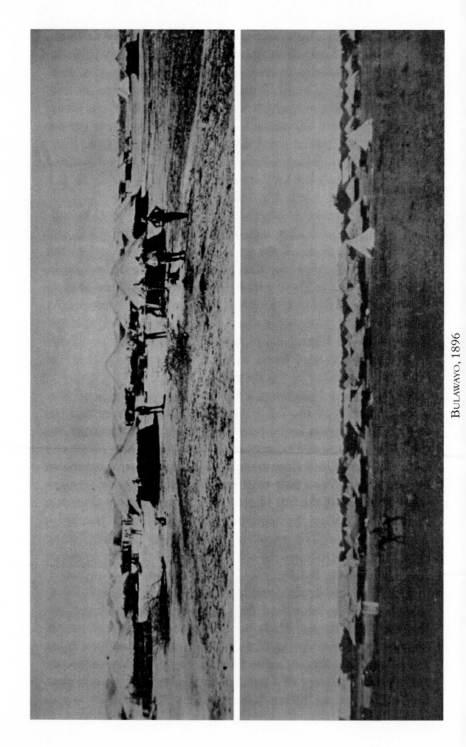

Bulawayo, 1896

command the column and Lieut.-Colonel Baden-Powell was to go as well. Dynamite and machine guns were also to be taken. All details not proceeding with the column were to remain at Gwelo with the stores, etc. The column was to travel as light as possible, and to return to Gwelo when the operations were concluded. A messenger therefore was immediately despatched to turn the column on to the Salisbury road, and Colonel Paget left Gwelo at 6 p.m. with rations for the whole column of two hundred men for ten days. Surgeon-Captain Hickson remained at Gwelo as he was suffering from fever. Dr. Brett accompanied the column.

Wednesday, 4 November.—Colonel Paget rejoined the column at Turf Flats, a place on the Salisbury road about twelve miles from Gwelo, at 10 a.m. The next day the column marched to Fort Gibbs at 4.30 a.m. The road was good and the distance covered twelve miles. Resuming the march at 4 p.m. another five miles was traversed.

Friday, 6 November.—Marching at 4.30 a.m. the column reached Iron Mine hill ten miles distant. Here Watson's column was found. Four sick men were left at the place in charge of Sergeant Shannon, Medical Staff Corps.

Saturday, 7 November.—Starting at 4.30 a.m. the column arrived at what was supposed to be Blinkwater, but this was not the case, as the real spot was two miles farther on. Length of march thirteen miles.

November 8.—Marched at the same hour to Sebakwe, a distance of ten miles. Resuming the march at 4 p.m., five more miles along a good road were covered. On the morrow Enkeldoorn Telegraph Office, twelve miles distant, was reached. Captain Eales. the Commandant of Enkeldoorn *laager*, rode out in the afternoon, bringing a sketch map of the position of the rebels on the Taba Inzimbi range.

Tuesday, 10 November.—Marched at 6 a.m. and formed a laager on the left of the road near the Enkeldoorn *laager*. At 4.30 p.m. A patrol of one hundred and sixty men with six wagons and the small Maxim gun started for Taba Inzimbi range. Having marched about three miles they halted for the night. Lieutenant Vaughan arrived with two wagon loads of grain which he had brought from the range, about seven miles from Enkeldoorn.

Wednesday, 11 November.— Marched at 4 a.m. to Hoffman's Farm, a distance of 8 miles. At 6.30 p.m. a patrol started for the position of the rebels, leaving the wagons with a small escort to follow at 3 a.m. on the morrow. About twenty Dutchmen arrived in camp at noon from Enkeldoorn laager, but as they refused to walk and brought no horses they were left to come on with the wagons, only the three Besters and Sergeant Cormack, who acted as interpreter, accompanied the patrol. The patrol halted at Bester's Farm about two miles from the poort.

Thursday, 12 November.—The patrol left the bivouac at 3 a.m. and advanced towards the rebels' position in the following order: Lieut.-Colonel Baden-Powell with two troops of the 7th Hussars proceeded to the east end of the range to cut off the retreat in that direction, taking one of the Besters with them as guide. The dismounted troop of the 7th Hussars and the mounted infantry (two sections dismounted) advanced under Lieut.-Colonel Paget and took up a position on the top of the range immediately to the east of the poort, and commanding the rebel position. The Maxim was taken with this party.

At first sight the place appeared to be deserted, but suddenly shouts and cries were heard and natives were seen running eastwards along the side of the hill on the opposite side of the valley. They had been startled by the appearance of Native Commissioner Taylor with his friendlies, who were advancing towards the *poort*. They did not seem to have discovered the presence of the rest of the patrol till firing began. Firing was shortly afterwards heard from the direction of the position which had been taken up by the two troops commanded by Lieut.-Colonel Baden-Powell, and Bester arrived with the news that the rebels had tried to escape in that direction, and that while so trying fifteen of them had been killed. The position was soon cleared of all rebels, and the *kraals* were destroyed by the *friendlies*, a small amount of grain and five head of cattle being captured. In the afternoon a troop under Lieutenant Imbert-Terry destroyed several large *kraals*. The patrol and the wagons started on the return march to Enkeldoorn at 4 p.m., and marched for two hours. There was heavy rain from 7 p.m. to 11 p.m.

Colonel Paget started at 11 p.m. to ride on to Enkeldoorn to meet Sir F. Carrington. The rest of the patrol was ordered to proceed direct to the telegraph station. Colonel Paget arrived at Enkeldoorn at 10 a.m. on 13 November and at once visited the camp of Carrington.

Saturday, 14 November, Lance-Corporal C. W. Barnes died of enteric in the hospital at Enkeldoorn and was buried near the *laager*. The patrol was found to have arrived at the telegraph office at 8 a.m. By orders, pending further instructions as to their leaving the country, the mounted infantry (2nd West Riding Regiment) under Captain Watson remained at Enkeldoorn, their strength being four officers, seventy-six N.C.O.s and men, ninety horses, six wagons and seventy-seven mules. The hussars marched for Bulawayo *via* Gwelo at 4 p.m.

15 November.—Marched at 5 a m. Arrived at Lebakwe at 9 o'clock. Resumed the march at 4 p.m. and covered four miles.

16 November.—Marched from 5 a.m. to 8.30 a.m. to Blinkwater. Resumed the march at 4 p.m.

7 November.—Marched at 5 a.m. to Iron Mine hill. Resumed the march at 4 p.m.

18 November.—Arrived at Fort Gibbs, having started at 5 a.m. and resumed march at 4 p.m.

19 November.—Turf Flats, starting the march and resuming it at the same hours. The column arrived at Gwelo at 9 a.m. on 20 November. Here rations were at once drawn and at 7 p.m. they started for Bulawayo. The column arrived at its destination on 28 November. No incident occurred on the march from 21 November till its arrival. At Bulawayo a standing camp was found to be provided. It consisted of a hut for each officer and one for every six men.

This is the last entry in the journal which has been placed at the disposal of the writer. That the account of the campaign as given is full of interest will be conceded. The details set down are clear and give hardships undergone by the troops engaged in this service. Another account of the Rhodesian Campaign of 1896-7, by Lieut.-Colonel R. M. Poore, D.S.O., is here inserted and is as follows:

At the beginning of May, three squadrons, A, C, and D, of the 7th Hussars (each nominally 100 strong) and 45 non-commissioned officers and men Royal Artillery, with 2 mountain guns, left Pietermaritzburg under Lieut.-Colonel H. Paget (7th Hussars), and travelled to Mafeking *via* Durban, East London (by sea), De Aar and Kimberley, the unsettled state of the Transvaal preventing any troops being sent through that State.

Two companies (100 each) of the York and Lancaster Regiment from Capetown joined this force at Mafeking.

The Jameson raid, which took place at the beginning of the year, had much disturbed the Boers and made them extremely apprehensive of any action the English were taking, and the force now collected at Mafeking, although known to be intended for operations in Matabeleland, was the subject of a good deal of correspondence between Pretoria and Capetown. Hence as the Boers seemed very suspicious, Sir Hercules Robinson, the Governor-General at Capetown, invited them to inspect this Field Force. Consequently, on the 8th June, President Kruger sent five Boer leaders headed by Commandant Schoeman who thoroughly inspected the force, after which they asked Colonel Paget to let them see the 40 guns which had been reported there, and though he denied their existence the Boer Leaders left the impression behind that they suspected this ordnance was hidden away somewhere. The Boer leaders lunched with the officers of the 7th Hussars before leaving.

In the middle of June the Field Force at Mafeking received orders to march north, on which it was decided to send A and D squadrons to the front, leaving C behind, and the first party left on the 22nd June.

Instructions were issued that each party should not consist of more than about 50 of all ranks (half a squadron) to facilitate the issue of forage on the line of march.

The railway extended north of Mafeking to Lobartsi, a distance of 40 miles, from which place marching commenced. Each party was supplied with 5 wagons drawn by spans of 12 mules with 2 spare.

Rinderpest had been very prevalent for the last six months, so much so that the country was practically denuded of oxen. Along the line of march thousands of these dead animals and hundreds of stranded wagons were to be seen.

The route travelled lay along the Western border of the Transvaal, and Boer piquets were constantly observed on the frontier watching the movements of the troops. Each party received instructions to march to Macloutsie a distance of

409 miles from Mafeking and situated near the N.W. frontier of the Transvaal, this march occupying about 22 days.

At Macloutsie fresh orders were received and the force was divided into two equal parties—the first to proceed to Victoria under Lieut.-Colonel H. Paget (7th Hussars), while the second was to go to Bulawayo under Major Ridley (7th Hussars), this latter consisting of D Squadron 7th Hussars, one Mounted Infantry Company of the York and Lancaster Regiment and 2 guns.

The rations for officers and men were as follows: 1 lb. meat, 3 oz. rice, 3½ oz. pepper, 1¾ lb. meal, 1 oz. Tea, ¼ pint lime juice, ½ oz. baking powder, ½ oz. Salt, 1 40th gal. *dop* brandy.

The journey on to Bulawayo was continued in parties of 50 as before. This road was much infested with lions, and one party had nearly all its horses stampeded by these animals and were fortunate in only losing four. A notice on the road side warned travellers not to let their animals stray on account of the lions.

This road ran through Mangwe where a fort had been built, and from that station to Bulawayo several forts protected the road. The march from Macloutsie to Bulawayo occupied about 13 days.

The chief officers in Matabeleland were: Lord Grey, Administrator, Sir Richard Martin. British Commissioner, Sir Frederick Carrington, Commanding troops in Matabeleland.

As soon as Ridley's column had collected near Bulawayo, orders arrived to proceed to the Gwai river. The column started on the 10th August, and after the first day's march a force of 800 friendly natives joined, armed with assegais and old guns. No wagons were allowed, each man taking one blanket. The troops lived on the country, for 10 days marching from 23 to 25 miles a day. and *en route* they accounted for about 150 of the enemy. The column returned to Bulawayo on the 20th August.

On the 26th August the column started North on an expedition to the Shangani river, passed Fort Inyati on the 29th and were joined here by 480 Volunteers.

Water in this country was very scarce and it had to be procured by digging in the river beds. Kraals frequently provided the column with grain which was collected for the horses.

The column arrived on the Shangani river on the 2nd September. On the 12th September Lieut.-Colonel R. S. S. Baden-Powell arrived in camp and took over command of the troops. His first act was to convene a General Court-Martial on the chief Uweena who had just been captured. The court sat after lunch on the 13th September. The accused was found guilty, the sentence was confirmed as soon as the case was finished, and the execution of the culprit took place the same afternoon. The old chief stood up most bravely before a detachment of volunteers who formed the firing party. His crime was that of murdering white settlers. Skeletons of white people who had been murdered were constantly found, some at their homes and others on the road.

Colonel Baden-Powell sent out three strong patrols, one of 20 under Major Ridley, one of 40 under Captain Poore and the third of M. I. under Captain Kekewich (York and Lancaster Regiment). All these started out on the 15th and 16th September, Colonel Baden-Powell himself accompanying Captain Poore's patrol, the first objective of this being the junction of the Umvungu and Shangani Rivers. Three days' grain had been taken for the horses, but no wagons, as it was intended to live on the country, *i.e.* to get grain in *kraals* for the horses and buck, &c. for the men, but no villages were found to provide the grain required, and rinderpest had destroyed the game. The result was that the horses soon became exhausted, and some had to be killed to provide food for the men. Lions also abounded, and precautions had to be resorted to to prevent them from taking the animals. On the evening of the 22nd a wagon arrived with provisions, and here Colonel Baden-Powell left this patrol, which eventually returned to Inyati with the remnants of their worn out and exhausted horses on October 3rd, arriving near Bulawayo on the 28th. Colonel Paget's and Major Ridley's column; having joined, the entire force marched into camp near Bulawayo on the 28th November. Previous to this the M. I. had been sent down country and the Matabele war was considered at an end.

A and D Squadrons remained encamped near Bulawayo from November till well into June the following year, and during this time they lost most of their horses from horse sickness which was very prevalent. In the meantime C Squadron had proceeded to Pietermaritzburg. where B Squadron had remained all the time.

About the 23rd June 1897 the two Squadrons of the 7th Hussars were sent North on different missions. D Squadron marched *via* Gwelo-Charter, and on the 12th July Matzwetzwe's stronghold was attacked at dawn, but its great strength was proof against an immediate capture. It was therefore invested and eventually taken. Major Ridley was among the wounded. Orders came that all the troops in Mashonaland were to concentrate at Mashigombi's stronghold by the 24th. On this day B Squadron joined A Squadron which had trekked *via* Gwelo and had then taken a more westerly road to that taken by D to the appointed place. A general attack was then commenced under the direction of Sir Richard Martin. The enemy were ensconced in fortified caves and were extremely difficult to get at, dynamite being required to be freely used to open up the caves. It took three days to capture the place.

On the 27th July the several units broke up in different directions, A Squadron under Captain Carew moving South to complete the work they had left unfinished and D Squadron under Captain Poore returning *via* Charter to Matzwetzwe, where a very large quantity of grain was collected and the fortifications destroyed.

On the 11th August A and D Squadrons united again at Salisbury, and after three days, halt moved out under the command of Major Ridley. They were accompanied by a large native contingent as well as some of the Mashonaland police, and the force operated in the Mazoe country, north of Salisbury, returning to the latter place on the 1st September.

The Mashona War was now considered at an end. During the halt that followed a number of horses died from the effects of eating the poisonous grass which is very plentiful in this country.

BULAWAYO, 1896

Arrangements were now made for the 7th Hussars to leave the country *via* Beira, and on the 23rd September they started on their march for Umtali. After proceeding for nine days, news was brought in that there was trouble to the southwest, so D Squadron was despatched in that direction. After marching over 40 miles, the offending Chief's people were found to have scattered, so the detachment returned to the road and resumed their march, reaching the vicinity of Umtali, near the Portuguese border, on the 7th. On the 15th all the horses were handed over to the Police, and next day the two squadrons marched out of Umtali on foot across the Portuguese border to Massi Kessi, which was then the rail head, a spot distant 201 miles from Beira. Here they entrained, and it being Portuguese territory they were escorted to the coast by a Portuguese officer. It is reported that the construction of this line caused a greater mortality *per cent*, *per* mile than even the Panama Canal. Beira was reached on the 18th and the troops immediately embarked on s.s. *Inyati,* the anchor being weighed the same afternoon. Durban was reached early on the 21st, and Pietermaritzburg the same afternoon, after an absence of 17 months.

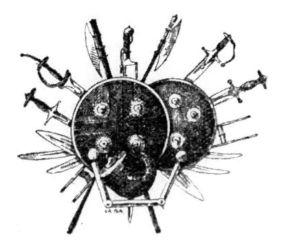

CHAPTER 9

Matabele
1896

We have now to endeavour to trace the movements of the detachment of the 7th Hussars which formed part of the command of Lieut.-Colonel Baden-Powell. Their junction with Colonel Paget's column at Gwelo has already been narrated. In Colonel Plumer's book *An Irregular Corps in Matabeleland,* page 180 we read:

> On the 13th August a detachment of 50 Mounted Infantry of the 2nd Battalion York and Lancaster Regiment came into camp; they were the first regular troops except the artillery to arrive.

He then goes on to narrate the various events which occurred. There is no mention whatever of the detachment of the 7th Hussars, so we know-not when they arrived. On page 196 we read as follows:

> The patrol, of detachments of the 7th Hussars and York and Lancaster and West Riding Regiments, under Colonel Baden Powell, went up to Inyati, through Tabas-i-Mhamba. which they found deserted, and on to the Shangani river. They came upon several small parties of rebels, whom they dispersed, captured the chief Uwini. worked through a great portion of the Somabula forest, captured the stronghold of the chief Wedza, and finally joined hands at Gwelo with the other party of the 7th Hussars and West Riding Regiment under Colonel Paget.

In the preceding chapter the adventures of Colonel Paget's column have been fully dealt with.

The *Manuscript Regimental Record* gives us no details on the subject. Some facts are, however, to be gathered from Colonel Baden-Powell's *Matabele Campaign*. On 7 September 1896 he records that he had been ordered to go and take charge of the column which was then under Major Ridley in the Somabula forest. He had been very ill but was recovering. At sunrise on 8 September he started from Bulawayo, accompanied by three troopers of Plumer's men as escort. Their names were Abrahamson, White, and Parkin. Resting for the night at Fynn's farm, the march was resumed at daybreak on the 9th. At 8 a.m. the party reached Inyati, where Imbert-Terry of the 7th Hussars was found, with six men, occupying a small fort. They were in charge of stores of food and grain.

That evening they reached the Longwe river, where they met with a convoy of four wagons with supplies for Ridley's column, but could obtain no information as to where the column was camped or how far ahead it might be.

The convoy had a strong escort. On n September Captain Vyvyan arrived from Major Ridley's camp. He was on his way to Bulawayo to act as Chief Staff Officer. Ridley's column was then twenty-five miles away. Colonel Baden-Powell reached the camp during the morning of the next day. He then took over the command of the column. Paying a visit to the hospital, he found there a man of the 7th Hussars, a noted football player, who had been wounded and whose hand had in consequence been amputated. Uwini, a chief, had been wounded and was a prisoner, but obstinately refused to allow anything to be done for him, invariably tearing off the dressings from his shoulder, through which he had been shot. It appears that the column had already lost five men in taking one *kopje*, and seven more *kopjes* remained to be taken.

The force commanded by Lieut.-Colonel Baden-Powell consisted of a squadron of the 7th Hussars under Captain Agnew, a company of the York and Lancaster Mounted Infantry under Captain Kekewich, a strong troop of the Africanders Corps under Captain van Niekerk, three Maxims, a seven-pounder under Captain Boggie, field hospital under Surgeon Lieut.-Colonel Gormley, ambulance and wagons carrying about a month's stores; a total of three hundred and sixty men and horses. The camp was situated close to the edge of the Somabula forest and on the bank of the Uvunkwe river, a river which joins the Shangani river about fifty miles farther on.

A court-martial was held on the chief Uwini. He was charged with armed rebellion and ordering his people to murder whites, and also for instigating rebellion. Uwini after a long trial, in which he practically acknowledged his guilt, was found guilty and sentenced to be shot. The execution duly took place at sundown on 13 September. That evening Major Ridley made a night march with fifty hussars to attack a party of Matabele who were camped about fifteen miles to the south on the Uvunkwe river. Shortly after leaving the camp, while passing near a piquet which guarded the enemy's water supply, they were mistaken for Matabele and fired upon. Luckily there were no casualties. The enemy, being unable to obtain water, were now very thirsty. Whenever they approached to get some they were fired at. The execution of Uwini, too, had had its effect, many in consequence slipping away and the rest surrendering, several with their arms. Large stores of grain were found at Uwini's stronghold when it was examined. Nearly one thousand of the enemy were prisoners in camp, and there was thus no lack of labourers to collect and store it. Major Ridley's patrol returned early in the morning, having found the *scherms* of the enemy deserted, and from the tracks the natives had cleared into the forest.

On 14 *September* Colonel Baden-Powell started against the Somabula forest. Captain Agnew was left in command of the remainder of the force in camp, to collect the grain and receive the surrender of rebels. The force on the patrol amounted to one hundred and sixty hussars and mounted infantry, with two guns, an ambulance, and four lightly loaded wagons. They marched in a northerly direction with the intention of breaking up the rebel *impi* reported to be located near, also to clear the forest, and finally to break up rebel posts which had been so placed on the main roads as to prevent rebels inclined to surrender from coming in. Before dawn on 15 September the patrol crossed the Uvunkwe river to the grain-fields and villages of Lozan. These had been quite recently deserted. A few women were captured: some portion had come from an *impi* in the Mashene hills, which it had been intended to attack. That *impi* was, however, reported to be already on the march to the north-west, owing to the approach of Colonel Paget's column from Victoria *via* Gwelo.

Colonel Baden-Powell therefore turned off to the North to intercept this *impi* if possible. Such a route would, however, have taken the column through a part of the forest which was full of the enemy, and he consequently determined to divide his column into three distinct patrols who moved rapidly through the forest on parallel lines, the wagons being left with a party to follow along the central route, thus forming a supporting depot and reserve.

The trouble now was lack of water; but this want cut both ways, as it equally affected both the British and the rebels. Hence it was only needful to search the existing water places, as thence the tracks would assuredly lead to the enemy. The patrol camped in the middle of the day on a drift where the Hartley Hill road crosses the Uvunkwe river. Close by they found the remains of three murdered white men.

The three patrols started in the evening. One, under Captain Kekewich, of thirty Mounted Infantry, was to proceed through the forest, and then to follow the course of the Gwelo river and to get into touch with Paget as soon as practicable, provided that the forest was thoroughly searched. The second, of forty men under Major Ridley, was to work along the Uvunkwe river, which flowed along the edge of the forest on the left. The third patrol, consisting of forty hussars under Captain Poore, with whom Lieut.-Colonel Baden-Powell also went, was to proceed rapidly down the Uvunkwe and through the forest to the lower part of the Gwelo river, to cut off any rebels retreating from either Paget or the other two patrols. From the lower Gwelo river they would be in position to reach a path which leads north, where the country was a great place of refuge for the rebels, where the grain-bearing district of Inyoka was situated.

The wagons, guns, and ambulance followed Major Ridley's patrol. About four days' supplies were carried by each man. On the first night eleven miles were covered, when the third patrol unsaddled and went into bivouac.

Starting on 16 September before dawn the patrol made its way along the river and found no traces of natives. Later they got into a very thorny tract of bush, which it was quite impossible to negotiate, and had to return to the river. Here a perfectly fresh track was found, which led from the water into another part of the bush which had just been found impracticable. The whole party dashed up the path and into the bush, where numerous huts and *scherms* were soon discovered. The rebels fled as the patrol rushed the place. Their fires were burning, their cooking pots were on, and arms, clothing, and loot were lying about in great quantities, all the spoils of luckless white men and women who had been murdered.

Leaving Captain Poore and the hussars to destroy the huts, Colonel Baden-Powell with three hussars followed the track for another three miles, till the sun had set and darkness came on. Nothing more was to be seen of the rebels, and the four then returned to the patrol, who had by this time gone into bivouac on the river's bank. As the rebels must come that way for water, fires were lighted all along the river bank after dark for nearly a mile opposite the spot where the bush joined the river. By this manoeuvre the enemy would be rendered afraid to approach, under the delusion that the force was much larger than in reality it was. After a meal the march was resumed in the darkness, and the patrol bivouacked for the night below the junction of the Uvunkwe and Shangani rivers.

From a small black-bead necklace which was found among the loot before mentioned it was concluded that a particular and important Matabele regiment known as M'tini's, which acted as bodyguard of a certain M'qwati, the high priest of the M'limo, was quite near, and on 17 September Colonel Baden-Powell started before dawn with a patrol of a dozen men to search for these rebels. Meanwhile Captain Poore took another patrol up the Shangani, with a view to intercepting any rebels who were retreating in that direction. Major Ridley's patrol was seen, and arrangements were made to cut off the water supply from both the rivers on the next night.

Lieut.-Colonel Baden-Powell succeeded in finding the huts but without capturing the rebels, though he came near to doing so. The huts were burnt. Twice more huts were found, but their occupants always managed to escape just in time. The patrol then returned along the to meet the wagons and obtain supplies of flour and coffee. After a rest the patrol started again to overtake Captain Poore's party, but before starting, in order to deceive the enemy, fired both in volleys and independently into the bush. The patrol after a toilsome day reached Captain Poore's men late that night.

Next morning they started as usual at 5.15 a.m., marching

northward along the Shangani river. They next struck across the forest to the Gwelo river, which was supposed to be about twenty-five miles away. About a mile from the Shangani river an unexpected stream was met with, where the spoor of the natives was hardly a day old. Here a woman was captured who informed them that a party of M'tini's *impi* was camped near. This was confirmed by another prisoner, a boy, who offered to guide them thither. He was put up in front on the horse of one of the hussars and soon led them to the *kraal*. The rebels were completely surprised, and being surrounded by the hussars with drawn swords, were captured.

A second party of rebels was similarly netted by a detached party. The prisoners were then taken to the water, where the patrol halted. They did not seem at all sorry to be captured, and the women built huts for their captors, while the children lit fires and boiled the kettles, and the men killed goats and cut them up. It was ascertained that they were tired of the war and anxious to surrender, but that their chiefs forbade it, and had placed piquets on the path to prevent them from giving themselves up.

They stated that the path (a new one) led direct to the Gwelo. Two prisoners were retained as guides, and the rest were told to go down to the wagons and report themselves as prisoners. This they were quite prepared to do. It was also learned that a large body of rebels was massed on the Gwelo river.

The patrol started in the evening, having first visited a ruined homestead which had been sighted earlier in the day. Here a murdered white man was found. The remains of the body were buried. The road through the Somabula forest was toilsome, the pace slow, and the way dark amid the trees. The hard work, too, and want of proper food told heavily against the horses. Grain they had not, and the forage obtainable was withered and parched. Water was scarce, the heat most oppressive, and the sand hot and heavy.

After a march of some hours one of the horses gave out, several were hardly able to move, and a halt became imperative. Finding a slightly open spot, the horses obtained what food they could and the men rested. The march was again renewed and continued till midnight. It was found impossible to reach the Gwelo river before dawn, and consequently all hopes of surprising the enemy had to be abandoned. The wearied horses and men—for they had marched forty miles—therefore halted, off-saddled and bivouacked.

The march was resumed at dawn on 19 September. After going three miles the edge of the forest was reached and the river Gwelo lay before them. No rebels were to be seen, though deserted scherms had been passed. The guide stated that the rebels were probably on a little stream about a day's march on across the Gwelo. While the drift was being examined, the fresh spoor of two men was found which led in a northerly direction. These were asserted by the guides to be those of two men who were hurrying on to give notice of our approach. The condition of the horses precluded a pursuit, and there was nothing for it but to return.

Accordingly the patrol retraced its steps, the hussars mostly walking the whole of the way to save their wearied horses, and despite the fact that their boots were in the last stage of dilapidation.

On 20 September the patrol marched at 5 a.m. and followed the course of the river for some miles, intending then to strike across the country. According to such maps as existed, this was possible to be impossible. As the heat of the day increased, so did the difficulties and sufferings of the horses and men. After going six miles it was found impossible to reach the Shangani that day.

Two horses had already been abandoned, and several completely done up. There was nothing for it but to return to the Gwelo.

As during the day the carcases of animals that had evidently died from rinderpest were discovered, cattle could neither be possibly obtained for food nor game shot. A horse was therefore killed and cut up and issued as rations to the men. The guides were now given as much food as they could carry and a supply of water, and despatched with a note to the wagons to ask for supplies on packhorses to be hurried up.

The patrol then turned back to the Gwelo river. They proceeded along the bank till a path was found southwards at the foot of a tall fruit tree, which path had been stated to lead quickly to the Shangani river. The patrol followed this path, every man being on foot and in Indian file. They marched till past midnight, but no river Shangani was to be found. A halt was called, and while the remainder of the patrol rested, Lieut.-Colonel Baden-Powell and an American Scout who was with the party, mounted on ponies, went on ahead to try and find the river. For nine miles the pair proceeded by moonlight, but the moon was by this time getting low and there was nothing for it but to return to the patrol, and moreover to again return to the Gwelo river. They had proceeded only a short distance towards the place where the patrol had been left when complete darkness set in, though dawn would break very shortly. A fire was made by the two wearied men and cocoa was cooked. With dawn Colonel Baden-Powell climbed a tree in hopes of sighting the rivet No water was, however, to be seen. As he descended he heard a noise in the bush, which proved to be the patrol, which had followed them up This decided the question as to whether they should press on or return: press on they must. The party was halted for rest, and meanwhile another effort was made to find water, this time by striking off the path at a spot where the ground seemed to slope downwards. After proceeding for some miles, the two decided that no more could be done, especially as the ponies were now nearly knocked up. So an attempt was made to rejoin the patrol. On the way back and quite by chance water was found. An hour later the patrol was at the spot and were off-saddled, camped, and watered. Provisions, however, were nearly exhausted, one pound of bread, a little tea, and a spoonful of rice remaining for each man. Salt, sugar, and coffee were entirely lacking, though plenty of horseflesh still remained. The American Scout and Corporal Spicer of the 7th Hussars were now mounted on the ponies and sent ahead to ride to the wagons as soon as the extreme heat of the day was over. At 4 p.m. the remainder of the patrol moved on to the south and east until long after dark. A halt was then called as two more horses had given out. Their saddles were taken off and transferred to other horses, some of whom were already laden with two or three.

Suddenly and quite close to the halting place the wide expanse of the river had appeared. Fires were lighted and the wearied men camped on a tree-shaded knoll overlooking the river.

On 22 September the march was resumed along the Shangani river for two hours. This day another horse was killed for food. Some of the hussars also managed to catch some fish in the pools of the river, thanks to the fact that a man had some fish-hooks with him. During the afternoon march Lieut.-Colonel Baden-Powell took a small patrol away on the flank of the main party. As they had not rejoined when darkness came on, signal shots were fired to show where they were. The patrol replied with three shots in rapid succession, as was usual. To the surprise of everybody this was followed by a volley and a signal fire.

The patrol made its way in that direction. More shots were fired and replied to, and on reaching the place the relief party under Captain De Moleyns, which had been sent out to meet them, was found in camp. Fires were burning, and a plentiful supply of food was laid out for issue, and a generous amount of forage for the famished men and horses.

On 23 September the main patrol under Captain Poore rested while a small party rode back with Lieut.-Colonel Baden-Powell and Captain De Moleyns to meet the wagons, a distance of twenty-two miles. After resting the main patrol followed by easy stages.

The objective was now Inyati, as an impi was reported to be collected in that neighbourhood. Inyati was reached on 1 October, and on arrival a letter was awaiting from Sir F. Carrington stating that the *impi* which they had come to search for had intimated its willingness to surrender. The patrol was, however, ordered to cooperate with Colonel Paget against Wedza. Accordingly the best horses of the hussars and Mounted Infantry were picked out to the number of one hundred and fifteen, and with a seven-pounder and two Maxims, accompanied by three weeks' provisions in wagons, a start was made to assist in the capture of Wedza's stronghold. Prince Alexander of Teck was now appointed Staff Officer to Lieut.-Colonel Baden-Powell and worked hard at the arrangements needful for a start on 2 October.

Early that morning the small column started; its strength consisted of half a squadron of the 7th Hussars, the York and Lancaster Mounted Infantry, together with the seven-pounder and the machine guns manned by police under Captain Boggie—about one hundred and sixty men, to which must be added the ambulance, and wagons carrying stores and supplies for three weeks.

The route taken was across the *veldt* in a south-easterly direction towards the Belongwe District. After various adventures the column arrived on 9 October in sight of Wedza's mountain and camped at Posselt's farm, where a fair amount of grazing was to be found. Here they were joined by twenty men of the Belongwe garrison under Lieutenant Yonge, and bringing with them a Nordenfelt gun.

On 12 October the column reached the Umchingwe river. From signal smoke-fires kindled by the natives it was clear that the strongholds of Monti and Matzetetza were occupied. Some patrols had been sent out, one to endeavour to locate the cattle of the rebel Wedza with a view to raiding it, the other to collect and bring in a force of friendly natives. That night all the available mounted men, one hundred and twenty in number, were ordered to proceed with two days' rations to reconnoitre Wedza's stronghold. Lieut.-Colonel Baden-Powell started during the afternoon with only an interpreter to endeavour to hold conversation with Matzetetza, and to persuade his people to surrender.

The *kraal* was, however, found to be deserted. They then returned and met the patrol on its march.

The road was bad; rivers, rocks, and boggy ground had to be crossed, and it was not until midnight that they reached a valley beneath one face of Wedza's mountain. Fires were burning both on the mountain and the opposite ridge. The patrol then bivouacked.

At 2.30 a.m. on the following morning, leaving the horses under a guard of fifty men, the remainder of the patrol went on foot to the base of the mountain, hoping to be able to ambuscade some of the rebels when they came down to get water, and to obtain information from any captives taken. The nature of the ground, however, was too difficult for them to reach the required position before daybreak. The men therefore concealed themselves. By a chance—a lucky one—a conversation was entered into with some of the rebels, Wedza himself being present. All efforts to persuade him to surrender were without effect, and that worthy merely suggested that if they wanted to capture his stronghold it was open to them to do so.

The presence of the British force being perfectly known now, the rebels watched them from secure spots in great numbers. Early in the afternoon the patrol moved off to attack the ridge opposite to the mountain and on the other side of the valley. Here four *kraals* were taken, and burnt and a few cattle were captured, as well as some goats and sheep. The natives did not show fight, but bolted and hid themselves in caves. The patrol returned to the valley when darkness came on, and again bivouacked.

Wedza's stronghold, which was by no means an easy place to take. now remained for the morrow.

The stronghold itself was a long six-peaked mountain, and strong *kraals* topped each of the peaks. The width of the position was about a mile and a half. There were other *kraals* on the side of the mountain, which was steep, rocky, and bush-clad. Small paths led to the *kraals* and these paths were fortified. Altogether it was a very hard nut to crack. First, however, it was determined to capture Matzetetza's stronghold, and this was undertaken on the morrow. To effect this the guns were sent for. During the heat of the day the force remained quiet. In the afternoon the guns arrived and were placed in position. The operations were plainly visible from Wedza's stronghold, and were calculated to produce considerable effect upon that potentate. After shelling the stockaded entrance and some of the caves, a strong party was sent up into the *kraal* with covering parties in case the rebels attempted any surprise. The enemy, however, did not fight; in fact, they had abandoned the place and their stores of grain, goats, and poultry earlier in the day.

After dark the operations were finished and the force camped by some water near the place.

A message was now received front Colonel Paget announcing that he would not be able to join in the attack on Wedza's stronghold. It was then determined to proceed against the place without further delay, though this, of course, necessitated a complete change in the plan of attack.

In the event the Mounted Infantry of the York and Lancaster Regiment under Lieutenant Thurnall, leaving their horses in the valley, took possession of a neck which joined Wedza's mountain to the northern range of mountains. Their orders were to hold this position for the whole of the day and the night, if not for longer. Their number was only twenty-five, but they were to make themselves appear as numerous as possible.

The guns were destined to bombard the centre of the position, and the left flank and rear of the rebels were to be threatened by parties of the 7th Hussars. Early in the morning the Mounted Infantry reached the northern end of Wedza's mountain, and here seven men were sent to seize a mountain which overlooked the position. This they did in about an hour, but did not gain the top before they were discovered by the enemy and a heavy fire was opened on them. Among the rocks at the top, too, a small body of the rebels had taken up a position, and it took some time before they could be dislodged. By this time the whole mountain was thoroughly aroused, and large bodies of the enemy could be seen apparently intent on attacking the seven men of the Mounted Infantry who held the mountain-top previously mentioned. A message was sent to the guns to open fire without delay, but it transpired that they had been detained by the wagons which were carrying away the grain from Matzetetza's captured stores. The seven men who held the horses were then hastily mounted, the remainder of the horses being left unattended. These with all haste worked round through the bush to the left rear of the stronghold, where there was a large village. This was at once attacked; the natives rushed away to caves for refuge; the alarm spread, and the rebels above, fearing a fresh attack in a new direction, collected together on the peaks with considerable excitement. This had the effect of diverting the rebels from their intention of attacking the Mounted Infantry. And now the hussars and artillery arrived, and by their arrival occasioned more terror to the rebels. The attacking force then rested for a brief time, and after the arduous work of the morning a rest was well earned. Presently the attack was resumed. The seven-pounder opened fire, and after two or three shells had been sent on their way, the rebels were observed to be on the move.

Here the little party of Mounted Infantry joined in and smote the shaken enemy whenever an opportunity occurred. This hand-ful of Mounted Infantry were now in great straits for want of water, and made signals to that effect. An attempt to carry water to them failed, the relieving party being driven back. They then signalled that numerous bodies of the rebels were escaping by a path which was out of their range.

Lieut.-Colonel Baden-Powell, with Mr. Jackson, the Native Commissioner, then rode round to the back of the mountain, di-recting Prince Alexander of Teck to bring on some of the 7th Hus-sars to the path in question. Having succeeded in reconnoitring the main paths towards the mountains to the north and east, Colonel Baden-Powell returned towards sunset, and by the way narrowly escaped being shot. The firing brought up Prince Alexander of Teck and the hussars, who arrived most opportunely, and speedily cleared and occupied the rocks where the rebels had lain in ambush. Prince Alexander then caused fires to be lighted round the flank and rear of the rebel position. Lieutenant Thurnall did the same on the heights north of the stronghold, and the men left in camp similarly along the front of the rebel centre. This gave the enemy the notion that a far larger force was attacking them than was really the case. During the night they fired volleys into isolated fires, but as the men were nowhere near them, by order, no harm was done.

Early on the morning of the 16th the seven-pounder and the machine guns opened fire upon each *kopje* and each *kraal* in succession. The rebels cleared from the *kraals* and took refuge in the caves. whence the seven-pounder shells soon drove them, and they slipped away through the rocks and bush. Wedza's own *kraal* was now taken, and here subsequently were found Matabele arms in numbers and a large store of dynamite. this last the whole *kopje* was blown up and destroyed.

The hussars were now recalled from their positions around the stronghold, and headed by Major Ridley, eagerly clambered up the mountain to join with the Mounted Infantry in completing the destruction of this hornet's nest. This laborious work—work accomplished throughout the burning heat of the day—was, however, at length finished, and shortly after dark they had again descended the mountain-side, leaving all the rebel habitations in flames.

On the night of 19 October a party of forty mounted men, hussars and Mounted Infantry, moved out in pursuit of Wedza's fugitive people. Many difficult peaks had to be climbed and many very rough passages and caves to be explored. Rebels were not, however, to be seen, but villages and *kraals* there were ordered to be destroyed by ire, and nightly bivouacs were of course the rule. All this time, rebels were lurking in their various hiding-places and watching for every movement of the men in perfect security.

The patrol returned to their old camping place on 22 October. There were neither fires, camp, nor provisions ready. A letter from Major Ridley hung on a post stating that he had been ordered to co-operate with Colonel Paget against Monogula near Gwelo, and had therefore gone at once in that direction. Hungry, the unfortunate patrol had perforce to follow. At the next camping ground another note was found stating that he had gone on a few miles. After a short rest, during which the last tea and meal were consumed, the patrol again started, and after a march of twelve more weary miles, many of them on foot, they arrived at the wagons and their supplies of much-needed food.

On 27 October the patrol joined Colonel Paget's column, after marching over awful hilly, stony, and dusty roads, passing amid hills studded with deserted and looted miners' wattle-and-daub huts— for this was a district of gold reefs. The remainder of the events which took place after the junction of Lieut.-Colonel Baden-Powell with the column under Colonel Paget have been narrated in the preceding chapter.

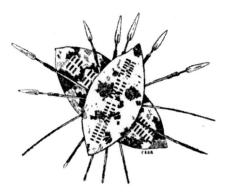

CHAPTER 10

Mashonaland
1897

We have narrated the course of the Matabeleland Campaigns in the two preceding chapters. Until June 1897 the two squadrons of the 7th Hussars remained in the standing camp at Bulawayo. Now, however, troubles began again, this time in Mashonaland. From the *Manuscript Regimental Record* we gather the following information:

> 1897, 22 June.—A native rising having taken place in Mashonaland, moved on June 22nd and 23rd to Charter and Salisbury, and from the latter place took part in the successful operations against the native chief Mashingombi.
> 18 October—The Mashona rebellion having been suppressed, the Regiment embarked at Beira on the s.s. *Inyati*, in which it proceeded to Natal, where it rejoined at Pieter Maritzburg the 2 squadrons that had not taken part in the Matabele and Mashona operations.

We will now endeavour to give a more detailed account of these operations, as the story of an arduous campaign of four months is surely worthy of being related at greater length.

The situation was this. With the Matabele, who had been completely defeated and punished severely, we were at peace. With the Mashonas it was otherwise. During the earlier months of the year 1897 outrages had more than once been committed by parties of this turbulent race. The Mashonas differed from the Matabele in this. The latter possessed powerful chiefs—or had possessed them and to the orders of those paramount chieftains they were in the habit of giving obedience. With the Mashonas chiefs of a powerful kind were non-existent. Every petty tribe acted independently, and if inclined to work mischief did so on its own initiative and in accordance with the instructions which it happened to receive from the petty chieftain who chanced to be at the head of it.

Colonel De Moleyns, who now commanded the newly raised police force, was in the field in April 1897, and on May 3, in Capetown his capture of Shangwe's *kraal* in Mashonaland without loss was reported This stronghold was a position which the Matabele in their raids against the Mashonas had always failed to take. Many natives escaped during the night, but they left their women and children behind them. Twenty head of cattle were captured and forty goats. Fifty-four natives were known to have been killed. One trooper was wounded. The Mashona Chief Amapila was slain. Colonel De Moleyns' force was reported to have suffered severely from fever. In the opinion of the white population the action of the Government was not sufficiently energetic; this was not, however, anything new.

On 3 June their wrath at the apathy of the ruling powers in dealing with the turbulent natives was expressed at a public meeting held at Fort Salisbury. It appears that the natives had even lifted cattle from the Government *kraals*, and were yet unpunished. Later in the month there was a fight in which the *kraal* of a Mashona chief named Kunzi was attacked. A trooper named George Irwin was so severely wounded on the occasion that he died of his wounds on 20 June.

The 7th Hussars appear to have moved up from Bulawayo on 22 June—the date, be it remembered, of the Diamond Jubilee—and 23 June.

Their first objective was Umtzewa's *kraal* near to Fort Charter. This place had already been once attacked, but had not been taken. A second attack upon it took place on 14 or 15 July, in which a detachment of the 7th Hussars under Captain Poore was engaged. The rebels had forty killed while the British suffered no loss. The Unyani district was now cleared of the rebels, who had deserted their *kraals* and had taken refuge in the bush country, seven miles to the south. Preparations were now made to attack Mashingombi, a most truculent chief, whose hostile attitude was one of the principal obstacles to the pacification of the country.

On 15 July it was reported by wire at Capetown that Major Ridley's column attacked M'Guilse on 7 July. One private of the 7th Hussars was killed, and Major Ridley was wounded in the leg. Major Ridley was sent to hospital at Salisbury, whereupon Captain Poore assumed command of the 7th Hussars. The projected attack on Mashingombi was to come off as soon as a column under Munro which was on its way should arrive. Fighting still proceeded at Fort Charter. A march against Mashingombi was arranged for 20 July; meanwhile a column of one hundred and twenty of the Vryburg Volunteers, with whom were twenty Basutos, started for Langeberg to cut off the retreat of the rebels in that direction.

On 19 July, two hundred police and one hundred and fifty natives left Fort Salisbury to attack Mashingombi. They were to join the 7th Hussars at the scene of the operations against the rebels.

Next day Sir Richard Martin and his staff left Fort Salisbury for Mashingombi's stronghold in order to assume command against the rebels. These active operations were expected to begin on the 22nd.

On 26 July news arrived that Sir Richard Martin, who went to Fort Martin, thence to direct the operations against Mashingombi had succeeded in his object. The *kraal* was captured by a combined movement of the police from Salisbury under Colonel De Moleyns, and the 7th Hussars under Captains Poore and Carew, whose columns had joined. Fighting was, however, still continuing in the neighbourhood of the captured *kraal*. On this occasion the 7th suffered no casualties; but a few days previously, at a spot some twenty miles distant, Private Edward Simms was killed and Private James Peters severely wounded. Sir Richard Martin in his telegram states that it is impossible to say what the loss of the enemy may be. 'as the country is one mass of rocks and caves.' He adds that he cannot speak too highly of the manner in which 'Commandant De Moleyns and Captains Poore and Carew timed their marches and brought their columns into action.' The troops after the capture of the *kraal*, encamped on the position.

From another telegram we learn that the capture of the *kraal* was effected by means of a successful rush in the early morning. The rebels were completely surprised and were chased into the caves. A telegram from Salisbury, dated 26 July, informs us as follows:

Colonel De Moleyns' column arrived at a point four miles

from Mashingombi's main *kraal* on Saturday night (July 24). The force waited until daylight, and yesterday morning advanced and attacked the stockade. Colonel De Moleyns led the assault with 25 white police and 70 men of the native contingent. After the capture of the stockade, all the reinforcements available were sent up. Nine women and children were taken prisoners. Two bullets made of solid gold were found after the fighting.

It appears that the rebels kept up a steady fire on the piquets of Colonel De Moleyns throughout the night of 24 July. Many, however, endeavoured to escape, but were shot in the attempt. Among the bodies found was that of Mashingombi the Chief. Between four hundred and five hundred prisoners were now in the hands of the British. Subsequently Captain Carew with the 7th Hussars occupied all the positions on Marlie's *kraal*, capturing there over one hundred prisoners and without suffering any loss.

The hostile operations against Mashingombi were eminently successful. Sir Richard Martin returned to Salisbury. The country each side of the road was everywhere patrolled. The 7th Hussars under Captain Carew moved in two columns down the river Unfuli to Charter. The whole of the Hartley district was now in the way to be quietly and securely settled.

On 23 July the death of a private of the 7th Hussars named Dands was reported, but the circumstances are not stated. Affairs remained in a state of quiescence until 9 August, when a despatch from Fort Salisbury announced that a patrol of the 7th Hussars and some police combined was about to start in search of the witch-doctors Kubube and Myanda. It was also stated that the petty chiefs of the Hunyani district were coming in with peace offerings, but that up to date their people had not followed that example.

August 13.—It was reported by wire from Fort Salisbury that an important engagement with the Mashonas at about thirty miles from that place was imminent. On 18 August, two Matabele found guilty of the murder of the Cunningham family at the beginning of the revolt were tried, convicted, and sentenced to death. August 28.—A wire was received stating that the Mashonas were surrendering satisfactorily. By 8 September all the Mashona chiefs except Kazubi had come in.

Magwendi and the sub-chiefs Umleva and Gazi surrendered. The whole tone of the natives was found to be changed, and they were quite anxious for peace. Patrols which were sent out found the strongholds and *kraals* deserted by the rebels, who were all in hiding in the hills in small parties.

The paramount chiefs Chewesmve, Kunzi, Makunbi and Masimbura had all sent in messages announcing their intention to surrender. This message was sent from Fort Salisbury on 8 September. A Salisbury telegram dated 15 September tells us that all the patrols had returned after nearly a month's absence. They had met with no serious resistance. The rebels had at last begun to comprehend that they would meet with lenient treatment if they surrendered and gave up their guns.

September 29.—News arrived that a patrol—presumably of police—under a Lieutenant Eilet attacked and punished the Mashona Chief Orewa, who had declined to surrender. Twenty rebels having been slain and eighty women and children captured. Orewa wisely yielded.

It was now announced that the hussars would embark at Beira on 20 October. The rebellion was at an end. Satisfactory reports received from all parts of Mashonaland. Small chiefs here and there might require to be dealt with, but all organised resistance on a large scale was over; and any serious fighting in the future was improbable. Police stations were established or were being established, and as an evidence of the feeling of security prospectors for mining operations again starting out.

November 5.—The following General Order was published by the General Officer Commanding the troops in South Africa:

> The two squadrons 7th Hussars under Major Ridley returned from active service in Rhodesia to Headquarters Natal on 22nd Oct. 1897; and Major-General Cox, Commanding in Natal, has reported that then-conduct on arrival shows that a high state of discipline has been maintained during their long absence of eighteen months.
>
> This commendation has been received with much gratification by the General Officer Commanding in South Africa, and he congratulates the Commanding Officer, Lieutenant-Colonel Harold Paget, C.B., and the officers, non-com-

missioned officers, and men of the Regiment, on the high repute, known by many reports, to have been maintained by these their comrades, both as fighting men, and as soldiers of character and conduct.

The next entry is dated 16 August 1898, when the Regiment marched to Ladysmith and took part in manoeuvres in Northern Natal, returning to Ladysmith on 8 September. On 14 September the Regiment was inspected by Major-General G. Cox, C.B., General Officer Commanding in Natal.

The term of service in South Africa was now about to terminate and orders were received to prepare for embarkation during October. The Regiment had embarked for India in November 1886, and the time had now arrived for it to return to England. From Regimental Sergeant-Major Bone of the 7th Hussars an account of the following interesting incidents has been received:

During the Mashonaland Campaign of 1897, a small party of A Squadron, 7th Hussars, and a few natives were detailed to fetch rations and provisions from Hartley Fort. This party, under the command of Sergeant-Major Handsley, had to cut a way for the wagon through the forest. After getting the provisions and returning by the same track, Sergeant-Major Handsley and Sergeant Shannon of the R.A.M.C. were riding on the wagon and Corporal Bone was riding behind. Handsley jumped off the wagon and went into the bush and the party went on. As Handsley did not turn up Corporal Bone and Private Pratt returned to try and find him, but failed.

A party from the squadron went out at night to fire volleys, thinking they might attract his attention, and rockets were sent up, which also failed. The squadron had to be in position the next day or next but one, so they were compelled to go away, leaving Handsley to his fate. The Regiment was two or three days attacking Mashongombi's stronghold, which ended in the death of the chief and complete surrender of his tribes.

The following day a party consisting of Sergeant Surrett (since deceased), Corporal Bone, and four privates were hurried back to Hartley Fort to try and find something of Handsley. They patrolled the ground where he disappeared, and were assisted by a good native scout named John Dusselboom, who could trace Handsley's

SOUTH AFRICA

footsteps as far as the river bed. This party camped in Hartley Fort one night, barricading their horses in an iron shed, the men themselves sleeping in the fort. General Sir Richard Martin was at the fort and said the party might give up all hope of finding Handsley alive as the district was absolutely swarming with lions and no man could live six or seven nights without falling a prey to them. As a matter of fact, a huge male lion lay in the cattle compound. He had been shot overnight by a B.S.A. policeman, but not before he had killed two mules and two donkeys.

The next day the party rejoined the squadron, which had taken up the same camping ground as occupied the day Handsley was lost.

The following day the whole squadron under Major Carew, Lieutenant Holford, and Lieutenant Vaughan (now Commandant Cavalry School, Netheravon), went out in skirmishing order and patrolled the whole country for miles around, and returned to camp without having found any signs of Handsley. About two hours afterwards, while the squadron were having a meal, who should walk in but Handsley himself, absolutely worn out, and off his head. He was hatless, boots worn out, carried in his hand a broken bottle with water, and tucked in his shirt were a few wood-apples.

Handsley himself could not account for losing himself, but said he must have turned the wrong way when he wanted to return to the track which the wagon had taken.

The Regiment was on active service in Mashonaland, and during September 1897, A Squadron had bivouacked about twenty-five miles from Salisbury, Rhodesia.

Corporal Bishop went out alone with an ordinary sporting gun to shoot guinea-fowl, and when about a mile from bivouac he walked through some very high rushes. He had a sort of presentiment that something was following him, and turning, found a huge male lion which, making a spring, knocked him down. Bishop had the presence of mind to put the muzzle against the lion's chest and pulled both triggers (one barrel being buckshot which knocked the lion senseless. Bishop ran into camp in a fearful state of nerves, and next morning a party, under Major Carew went out and found the lion in the riverbed very much alive, and he was finished by a Martini bullet through his head. Bishop's helmet, which the lion had pulled off, was found near the beast torn to fragments. The skin of this lion, with a silver plate suitably inscribed, is now preserved in the mess.

CHAPTER 11

The Boer War
1900

On 29 October the Regiment, which had proceeded to Durban from Pieter Maritzburg, embarked there on board her Majesty's transport *Simla* for England. The vessel arrived at Southampton on 29 November, when the Regiment disembarked and proceeded at once by train to Norwich. Here they remained until 10 February 1899, when a detachment, under the command of Captain the Hon. R. Marsham, marched to Colchester to be quartered there.

St John's Gate, Colchester

The remainder of the Regiment followed by march route on 1 May, and having arrived at Colchester encamped on the Abbey Field in tents for the summer. The Abbey Field is approached by the gatehouse of the now vanished Abbey of St. John, Colchester, and is Government property.

This gatehouse is the only architectural relic of the Abbey which remains. At the dissolution of the monasteries it was granted to one of the Darcy family. Having soon fallen into ruin its stones were used to build a mansion for the Lucas family, who became the proprietors by purchase. This house was sacked, burned and destroyed, in the siege of Colchester during the Great Rebellion. The gatehouse itself was damaged by shot, the marks remaining to this day in the groining of the gateway. After the siege, Sir Charles Lucas the owner, was, with Sir George Lisle and a foreigner, selected by Fairfax for death. The foreigner was spared, fearing complications with his Government, but Lucas and Lisle were barbarously shot in Colchester Castle garth. They lie buried in St. Giles' Church, Colchester. The door of the cell in which these luckless gentlemen were for a brief time imprisoned is in one corner of the now roofless and floorless keep. The gatehouse itself has been restored in recent years. It is curious in Colchester to find the relics of the siege so plain and so numerous. One church tower is certainly still in ruins, and elsewhere the damage can be easily detected and shot marks discerned, though not so plainly as on

DUNGEON DOOR, COLCHESTER CASTLE

one of the church towers of Devizes in Wiltshire, where the hemispherical dints still liberally pit the old grey tower.

On 8 May the Commander-in-Chief, Lord Wolseley, inspected the recruits and young horses.

The term of command of Lieut.-Colonel Harold Paget, C.B., having now expired, Lieut.-Colonel the Hon. Richard Thompson Lawley was appointed as his successor on 26 June.

31 July.—Major-General Grant, C.B., Inspector-General of Cavalry, inspected the Regiment.

Leaving a detachment at Colchester as before, the remainder of the 7th Hussars returned to Norwich by train on 28 September.

The barracks, stables, institutes, &c. were inspected on 17 February 1900, by Major-General Abadie, C.B., General Officer Commanding the Eastern District. The spring inspection of outposts, riding school &c., which occupied 26 and 27 April, was made by Major-General Grant, C.B., Inspector-General of Cavalry.

On 1 May the 7th Hussars proceeded by train to Aldershot and were quartered in the South Cavalry Barracks. They were inspected on the 14th by Major-General Montgomery Moore, General Officer Commanding the Aldershot District.

By an army order (No. 173) of August 1900, the Establishment of the Regiment was fixed as follows: 1 lieut.-colonel, 5 majors, 5 captains, 10 lieutenants, 8 second lieutenants, 1 adjutant, 1 riding master, 1 quartermaster; Total, 32 officers. 1 regimental sergeant-major, 1 band master; Total, 2 warrant officers. 1 quartermaster sergeant, 1 farrier quartermaster sergeant, 1 sergeant-major rough rider, 1 sergeant instructor in fencing. 1 orderly room sergeant, 5 squadron sergeants-major, 5 squadron quartermaster sergeants, 1 sergeant trumpeter 1 sergeant saddler, 1 sergeant cook, 1 sergeant master tailor, 5 farrier sergeants, 37 sergeants, 1 orderly room clerk; Total, 62 sergeants.

10 trumpeters, 38 corporals, 4 shoeing smith corporals, 14 shoeing smiths, 5 saddlers, 1 saddle tree maker, and 720 privates. Total rank and file, 782. Total of all ranks, 888. Total horses (including transport animals), 601.

The summer inspection of the Regiment by Major-General Grant, C.B., the Inspector-General of Cavalry, was held on 20 August.

From 17 to 22 September an extended brigade reconnaissance

was held in the district around Frensham and Woolmer Forest, in which the Regiment took part.

On 12 November, twenty-five men under the command of Lieutenant Viscount Cole sailed for Australia, being selected to form a part of the force that represented the army at the inauguration of the new Commonwealth.

On 6 December the Regiment was called upon to furnish a draft of eight non-commissioned officers and men to join the 20th Hussars, who were then serving in India.

Meanwhile in October 1899 the South African war had broken out and from that time until the end of the year no fewer than three hundred and ninety-five horses were taken from the Regiment and transferred to cavalry regiments at the seat of war. The reserves were mobilised, two hundred and forty men of the 7th Hussars rejoining the Regiment in consequence. Of these two hundred and forty reservists, one hundred and forty were transferred to the 14th Hussars.

The following officers were also employed in South Africa on special service: Major R. L. Walter, second in command Bethune's Mounted Infantry. Major G. A. Carew, D.S.O., A.A.G., to the Rhodesian Field Force. Major J. S. Nicholson, D.S.O., Commandant Rhodesian Police, and subsequently C.S.O. Transvaal Police. Major G. L. Holdsworth, second in command Rhodesian Police, and afterwards CO. Bushman's Brigade. Major D. Haig, C.S.O. Cavalry Division, and afterwards in command of a column in Cape Colony. Captain Fitz Henry, Brigade Major, General Clements, Infantry Brigade. Brevet-Major R. M. Poore, Provost Marshal, Headquarters Staff. Captain R. G. Brooke, D.S.O., A.D.C. to General Sir George White and afterwards second in command of the South African Light Horse. Captain the Hon. R. H. Marsham, employed on Remount Duty in America. Captain H. Fielden, Adjutant 1st Regiment, Brabant's Horse. Captain J. Vaughan, Brigade Major 1st Cavalry Brigade and subsequently D.A.A.G. for Intelligence, Cavalry Division. Captain F. W. Wormald, A.D.C. to General Gordon, General Officer Commanding the 2nd Cavalry Brigade. Captain C. H. Rankin, Adjutant Rimington's Guides. Captain H.S.H. Prince Alexander of Teck, attached to the Inniskilling Dragoons and afterwards A.D.C. to Brigadier-General Mahon.

On 3 January 1901 the Regiment paraded dismounted to assist in lining the streets on the occasion of the return of Marshal Lord Roberts from South Africa.

On the occasion of the funeral of Her Most Gracious Majesty Queen Victoria the Regiment paraded in London. Three squadrons assisted in lining the streets while one squadron took part in the procession.

15 March.—Sixty-two non-commissioned officers and men sailed for India, being transferred to the 20th Hussars. These were followed by a further draft of seventy-six men, who were transferred to the same regiment and sailed for India from Southampton on 8 October in H. M. transport *Plassy.* The Regiment was now equipped with the Lee-Enfield rifle prior to embarking for South Africa. During the war the rifle was carried on the near side in a *Namaqua* bucket and attached to the man's left arm by the sling, the sword being on the off side.

On 1 November the Regiment received orders to prepare to embark for active service in South Africa. It was inspected on 21 November by Field-Marshal Lord Roberts, the Commander-in-Chief; and accordingly embarked at Southampton on 30 November, partly in H.M. transport *Templemore,* and the remainder three days later in H.M transport *Manchester Merchant.* Twenty officers, four hundred and nine non-commissioned officers and men, and four hundred and twenty troop-horses sailed in the *Templemore.* The names of the officers were Lieut.-Colonel the Hon. R. T. Lawley, in command; Major R. L. Walter, second in command; Major D. S. Carew, D.S.O.; Captains Norton, Dalgety. Fryer. Vaughan, D.S.O., and Johnstone; Lieutenants Leyland, Royds. Gibbs, Mann-Thompson, Gooch, Hermon, Henderson, Paget-Tomlinson, and Cooper; Captain and Adjutant Wormald, D.S.O.; Captain and Quartermaster Durman.

In the *Manchester Merchant,* which sailed from the Albert Docks, there were Captains and Brevet-Majors Brooke, D.S.O., and Harley. D.S.O.; Lieutenants Viscount Cole, Robarts. Kevill-Davies and Whitehead; with eighty-nine non-commissioned officers and men, and sixty-five troop-horses; these formed half of B Squadron.

Five hundred and twenty-nine non-commissioned officers and men and two hundred and sixteen horses remained at Aldershot, and with them were Captain the Hon. R. H. Marsham, Sir K. A. Fraser, Bart., H. Fielden, D.S.O., A. Imbert-Terry, Holford, D.S.O., H.S.H. P Alexander of Teck, D.S.O., K.C.V.O. ; Lieutenants H.R.H. Prince Arthur of Connaught, K.G., and F. Kelly, and Captain and Riding-Master Dibble.

The *Templemore* arrived at Las Palmas about 8 a.m. on 5 December coaled, and again sailed at 6 p.m. The vessel reached Capetown at 5 a.m. on 20 December. During the voyage the Regiment lost sixteen horses—fifteen from pleuro-pneumonia and one from inflammation of the outer casing of the brain. Six horses belonged to A Squadron, four to B, and the remainder, which included three chargers—those of Lieut.-Colonel Lawley, Major Harley and Captain Wormald—belonged to C.

The Regiment disembarked at 11.30 a.m. and marched to Green Point, where they went into camp.

The *Manchester Merchant* arrived at Capetown on 22 December.

On Saturday, 21 December, a few men left for Stellenbosch at 6.30 p.m. Next day C Squadron entrained for De Aar and two horses died on the way. The remainder of the Regiment having reached De Aar left it again at noon on 24 December. They passed through Wellington the same evening, where a halt was made to feed and water the horses. The train stopped early on the morning of Christmas Day at Tows river, later at Matjesfontein, where they met Colonel Haig, and next at Prince Albert Road, at all of which places either the men or horses were fed. The train arrived early the next morning at Victoria Road, and finally reached De Aar. Here they detrained, marched about a quarter of a mile, and then went into camp.

On the night of Tuesday, 31 December, at De Aar a very unfortunate accident occurred. A horse in the lines was shot by a veterinary surgeon at 10 p.m. The report caused a stampede. Practically all the horses in the regiment broke loose and the tents that were pitched between the lines were knocked flat, in many cases their occupants being hurt. A number of horses were killed and many more seriously damaged. Some broke their backs or necks, others were shot as they galloped along the line of blockhouses, and almost all the remainder were badly cut about by the barbed wire that surrounded the camp. A trumpeter was ordered to sound 'feed,' and many horses hearing the call came back, but it was several days before the sound ones were gathered in; seven were found at Britstown, a place thirty miles distant from De Aar.

On Friday, 10 January 1902, fifty remounts arrived for the Regiment.

On 13 January the Regiment marched out of De Aar in a south-westerly direction and camped that night at Rietfontein. The next night they camped at Taibosch. On 15 January they reached Hannover road on the Sea Cow river and camped. On the morrow the march was deflected to the north-east. They proceeded about twenty miles along the Sea Cow river as far as Klip Kraal.

On Friday, 17 January, they arrived at Colesberg Junction and there remained the next day. Six men were sent back from Colesberg to De Aar to go through a course of signalling.

On Sunday, 19 January, the march was resumed, and having camped that night, on the morrow they arrived at Norvals Pont. Here the Regiment entrained for Winburg at midnight on 21 January. And arrived at their destination in twenty-four hours.

At Winburg they joined the Queen's Bays. The weather was very bad. A column was now formed at Winburg which a consisted of the Queen's Bays, the 7th Hussars, two guns of the 39th Royal Field Artillery and a pom-pom; the command being vested in Lieut-Colonel Lawley of the 7th Hussars. The column moved out on Thursday, 30 January, at 4 a.m. towards Senekal, which lay to the east-north-east, in order to in that district of cattle and provisions. That night they camped at Reitspruit.

Next day clearing operations were continued. The column returned to camp at 3 p.m. An hour later a wire was received recalling them to Winburg, and they accordingly started, taking with them captured cattle and grain, and arrived at their destination at 9.30 p.m.

On Sunday, 2 February, at Winburg. a man was tried for refusing to obey an order. The sentence of the Field General Court-Martial on his conviction was two years' imprisonment with hard labour.

4 February.—The column left Winburg at 3 a.m. and encamped at the Doornberg on the Zand river, whence it was proposed to sally forth to clear the country of cattle and provisions. On arrival at Doornberg the advanced guard was sniped by a party of Boers some fifty in number. One or two horses were killed and the hat

of one of the privates was knocked off. On the pom-pom being brought up into action the enemy soon left.

Wednesday, 5 February.—A patrol of the Regiment met with a mishap on this day, being surprised by a party of Boers and captured. They were stripped of all their clothing by the enemy and sent empty away. Next day a similar misfortune occurred, only of a more serious nature, as one man, Private Burke of C Squadron, was killed, and two other men wounded, one in the calf of the leg and the other in the ankle. On this day a column under Major Ducane marched in and camped on the other side of the Zand river.

Friday, 7 February.—The column, consisting of the Queen's Bays, the 7th Hussars, two 15-pounder guns and the pom-pom, escorted Ducane's column to a new camping ground on the Kool Spruit and then returned to camp. The column left camp on the next day at 8.30 a.m., and after picking up Ducane's column started on a two-days' trek in a northerly direction, some Boers being reported to be in that quarter. That night the column camped at Potgeiters, Ducane's command camping at Welkomspruit. A small encounter with about one hundred of the enemy took place on 9 February: there were no casualties. That night they outspanned at Bloemhoek, and returned to their former camp at the Doornberg on Monday, 10 February. As regards health, up to this date there had been three cases of enteric. The weather was exceptionally bad; nor was this inclemency confined to the Doornberg. Elsewhere it is recorded that the guides refused to be responsible, owing to the heavy rains and thick mists.

On Wednesday, 12 February; all sick men and horses were sent in to Winburg. A drive was now projected up to the blockhouses, along the Kaffir Kop-Bethlehem line, and the column accordingly left the Doornberg on 13 February in order to take part in it. They marched in an east-north-east direction. The column had Ducane's men on the right, and on the left Marshall, Elliott, and Holmes.

That night they camped at Kleinfontein on the Kool Spruit.

On Friday, 14 February, the force started again at 6 a.m. A large number of sheep were met with on the road, and in one place four hundred were killed to prevent them falling into the hands of the enemy least twice as many, however, were left behind, as time would not allow of them being slaughtered. Nothing else of importance happened. They halted for the night at Pandam. An attempt made by the Boers to break through during the darkness failed.

Saturday, 15 February.—The column moved on to a place called Vaalkopjes on the Zand over, and by the traces of recent fires seen on the march it was obvious that the Boers were at no great distance ahead. During that night the enemy made a determined attempt to break through and there was a great deal of firing. When daylight came a number of horses mere found to be dead in front of the line. A corporal of the 7th Hussars was killed and Trooper Craigie wounded in the wrist, while another man was also hit. The drive was continued on 16 February up to the blockhouse line, but the enemy had as usual managed to sip away and none were captured by the 7th Hussars or in their immediate neighbourhood. That night the column rested at Kaffir Kop. Up to date the total bag was seven Boers, some cattle and horses.

17 February.—The column moved to Naude's Kop and joined Colonel Barker's column. Here they remained in camp until the 19th. The column started again at 6 a.m. and were accompanied by the columns of Heath, Marshall, Ducane, and Holmes, the whole being under the command of Colonel Barker. The country was cleared as the line advanced in a direction due east. That night they camped at Saldanka.

Thursday, 20 February.—The column moved to Tigers Kloof Spruit and camped. During the night Colonels Marshall and Holmes marched due north, to be followed by the remainder of the column on the next day. This move evidently took the enemy by surprise, and eighteen prisoners were captured at a farm. A number of cattle and sheep were also rounded up. These had evidently been driven hither by the Boers when the columns had moved east. That night the 7th Hussars camped at Bultfontein and spent the greater part of the time there slaughtering the captured sheep. A terrific thunderstorm took place during the night, accompanied by torrents of rain. Next morning the march was resumed at 10 a.m., this time to the north-west. The column arrived at a place called Grootfontein. When it became dusk they left on a night march to round-up some farms.

About 4 a.m. on Sunday, 23 February, they came upon twelve Boers sleeping in a cattle *kraal* with their horses close at hand. They were taken completely by surprise by two troops of A Squadron under Captain Wormald. Most of the enemy fled over the wall into a neighbouring garden. One young man was killed by a sword cut during the attack, and another severely wounded. Two who were with him escaped, but the rest, finding themselves surrounded in the garden and all chance of escape gone, surrendered at 8 o'clock with their arms or some of them: five rifles, one revolver, and a bandolier. The 7th Hussars remained at the farm till they were joined by their convoy at 2 p.m. and then moved on to Swartfontein, whence communication by lamp was established with Lindley.

Next day the column arrived at Lindley and obtained supplies at the Quaggafontein supply depot. On the morrow they moved out ten miles and camped at Plesier. The column was now acting as stops in a big drive, at the end of which one hundred and fifty Boers were captured near Harrismith.

Wednesday, 26 February.—The column left camp at 5.30 a.m., and having halted for three hours at noon at Elands Hoek on the Liebensberg's Vlei river, they moved on to Blignaut's Rust and then camped. They were now in sight of the Platberg (Harrismith), but owing to the mist it was impossible to establish communication by signal.

On Thursday, 27 February, they moved on to the Wilge river and camped. All the next day they were trying to communicate with Harrismith, but without any success. However, during a brief twenty-minutes' spell of sunshine a flash from a heliograph was seen at Tafel Kop near Frankfort, and that night they were in communication with Tafel Kop by lamp.

On Saturday, 1 March, a message came through from Lord Kitchener *via* Tafel Kop from Ladysmith, ordering the column to move south to Tiger's Kloof at once.

They accordingly left at 1 p.m. in a heavy thunderstorm and trekked to Roodepoort on the Leuw Spruit, where they camped for the night. Next day they reached Tiger's Kloof, having halted *en route* for three hours at Vrede.

During the rest of Sunday; 2 March, and the following night, they remained at Tiger's Kloof, preparatory to taking up their position for a drive towards the Vrede-Frankfort blockhouse line.

On Monday, 3 March, they left camp at 4 p.m. and moved into

position for the drive. They camped for the night at Heuining Laagte. Moving out at 6 a.m. on the morrow, the columns taking part in the drive were thus disposed: General Elliott on the extreme left, and then in succession Colonel Barker, Colonel Lawley, Colonel Rimington, and Colonel Kerr. The Wilge river divided the columns; as the force was acting an both banks. They halted at midday at Vrierfontein, whence a number of Boers and some cattle were sighted immediately ahead. Major Ducane had shelled them just before the halt. That night the column camped at Mahashi on the Leuw Spruit, and during the night there was a certain amount of firing, but it was probably chiefly due to the presence of some loose horses and cattle that had strayed to the front of the line.

Wednesday, 5 March.—The column started at 6 a.m. and marched in a northerly direction till noon, when it halted for two hours at Stoltskop. That night it camped at Vrede, except the A Squadron, which marched to Rooikraus, about three miles to the east, and camped there. During the night there was a good deal of firing, but it is doubtful whether it was caused by the Boers. In front of the main camp a Kaffir woman was killed; she had wandered into the driving line. Next day the column moved out at 6 a.m., and marched without halting until the blockhouse line was reached. The enemy, however, managed mostly to escape between the right of Rimington's column and the blockhouse line. Ducane captured eight prisoners, and Rimington seventy-five, as well as a Maxim and a quantity of Martini ammunition. That night the 7th Hussars camped at Driehoek. The convoy, which had been compelled to take a very circuitous route, only got in at 8 p.m.

During Friday, 7 March, the column remained quiet until 5.30 p.m., and then only proceeded a short distance to Dundas on the Groot Spruit. The wagons had been previously sent to Frankfort to load up supplies and only arrived in camp at 11 p.m. Here the column remained until Sunday, 9 March, when in accordance with orders received it moved out at 6 a.m. The intention was to proceed to Heidelberg to refit. Having marched *via* Herderdal to Alleman's Home, a spot south of Villiersdorp, a halt was made to water and feed. During the afternoon signalling communication was established with Tafel Kop, and there was also a certain amount of sniping.

On Monday, 10 March, an attempt was made to trap the Boers who had been at Villiersdorp and in the neighbourhood of the camp on the previous night. To effect this the 7th Hussars moved out at 4 a.m. and crossed the river Vaal to the left of the spot selected for the crossing of the main body. The idea was to get the Boers between the 7th and the main body: the latter only starting two hours later.

Unfortunately the scheme proved unsuccessful. A few Boers were encountered by the 7th Hussars on the left and some shots were exchanged without casualties on either side. The main body when they left the camp were pressed by the Boers from the rear for some time during their march through Villiersdorp, and when across the Vaal river a rearguard action was fought. The enemy were then shelled with the 15-pounders and the pom-pom, but it was impossible to see with what result, though the practice was apparently good. The only casualty was a Guardsman who was wounded in the foot. That night the column camped at Bier Laagte on the Kalk Spruit. Moving out at 6 a.m. on the morrow, the column halted to water and feed at Modderfontein, and resumed the march later, till at night the Natal-Heidelberg line was struck at Vlakfontein, where they camped. By noon on the morrow they marched into Heidelberg, where they remained for two days.

On Friday, 14 March, the column arrived at Springs, passing *en route* the Nigel gold mines and the Springs and Clydesdale collieries. At this juncture Colonel Lawley's staff was composed as follows: Captain Wormald, Staff Officer; Major Vaughan, Intelligence Officer; Major Smithe (Queen's Bays), Provost Marshal; Lieutenant Royds, A.D.C.; and Lieutenant Paget-Tomlinson, Signalling Officer. The staff was distinct from the Regiment and was quartered at Springs, two miles distant from the camp.

The column, consisting of the Queen's Bays under Fanshawe, the 7th Hussars under Walter, one section of artillery and a pom-pom, here took the place of Hamilton's column, which had been composed of the Royal Scots Greys and the 5th Dragoon Guards. The former regiment was leaving for Pretoria and the latter were under orders for India.

On Saturday evening the column, under Lawley, marched at 8 p.m. to Nigel. The intention was to occupy the Nigel mine under the cover of darkness, and to remain concealed in the buildings during the whole of Sunday, thence to march out on Sunday night and surprise a Boer laager which had been reported to be in the neighbourhood. The mine was occupied at 1 a.m. on Sunday and the men remained under cover throughout the day. Scouts were, however, sent out, and these reported that the Boers had cleared; so the plan proved abortive.

The column, nevertheless, marched out that evening to a place called Nooigedacht, a distance of fifty miles, where the enemy was reported several times to have *laagered*. The column arrived at 2.45 a.m. on Monday morning, but found no signs of the enemy. They off-saddled for an hour and then marched back towards Springs. At Vischkuil they halted for two hours for breakfast, and then proceeding arrived at their destination at noon on March 24 The column remained quiet for three days, when C Squadron was sent out to reconnoitre, but returned on the following morning, having marched forty miles and achieved no success. About this time a fresh lot of remounts arrived for the regiment from Elandsfontein, to the improvement of its efficiency, as the animals were of a decent quality. One of the remounts, however, kicked Sergeant Moine in the stomach on the 27th, and the unfortunate man died from his injury on the 28th.

On 29 March the column left at 2 p.m. for Boschmanskop, a place distant about sixteen miles from Springs. They arrived there at 7 p.m., and having off-saddled and fed awaited the rising of the moon. They then saddled up, and leaving the wagons behind protected by the guns, made a night march to Strekla, when it was reported that the Boers had collected to pray for peace. They reached the spot just as the sun rose, but were again disappointed. After an hour's rest the return journey began, and just before halting for breakfast three of the enemy were started from a farm. Two escaped, but the third was captured. From the prisoner it was learnt that there were one thousand two hundred Boers close at hand, but the horses were already done up, and so it was decided not to attack them. After a halt of two hours the column started again and reached Boschmanskop at tea time.

Monday, 31 March, was spent in camp until the evening, when the Bays made a night march to Holspruit, a place near Leuw Kop. There they engaged the enemy about an hour before dawn. They were greatly outnumbered by the Boers, but in spite of heavy odds acquitted themselves well. They fought a rearguard action back to Boschman's Kop, when they were reinforced by the 7th Hussars—who had turned out as they came in sight—supported by the fire of the guns and pom-pom. The Bays lost heavily, especially in officers. Out of seven officers, two were killed and two wounded, one of whom was Vaughan of the 7th Hussars. The whole of Tuesday was in consequence spent in bringing in and attending to the wounded, who were sent on to Springs the same day, and from thence to Elandsfontein. Another account tells us a few more details. The commando of Pretorius was the first of the enemy that came into action. They were followed by those of General Alberts and others. The Bays were forced to retire on Boschman's Kop. About 6 p.m. they were reinforced by the 7th Hussars and the enemy were then driven back. The casualties were as follows: Queen's Bays, two officers and thirteen men killed, two officers and fifty-nine men wounded; 7th Hussars, two officers wounded. The loss of the enemy was about thirty killed and eighty wounded. As will be seen, there is a discrepancy here in the number of casualties.

On Wednesday, 2 April, more wounded were brought into camp and were sent next day to Springs. That evening a convoy escorted by South African Constabulary arrived, carrying provisions for some of their people who were establishing a line of posts along an intended blockhouse line. Three posts were being established on the kop, and the column remained till these were secure.

On Thursday, 3 April, the column was in communication with Springs and Eden Kop, both by lamp and heliograph. Next day they returned to Springs to refit, and remained there until 8 April. They then marched at 7 a.m. to occupy Leuw Kop, to enable the South African Constabulary to establish a post there before building a blockhouse. That night the column camped at Boschman's Kop.

Next day they reached Leuw Kop without meeting with any opposition, and camped that night close to the place. A farm was captured during the day where two Boers were found. One was a wounded Boer lieutenant and the other was taking care of him. They belonged to Naude's commando. Nine rifles and four thousand five hundred rounds of Martini ammunition were also captured. That night a medical man was sent in by the enemy to ask for the wounded man. His request was not, however, granted, and he was moreover detained until mid-day on the 10th.

The column left camp at 6.30 a.m. and trekked *via* Rolspruit (where they halted two hours for breakfast) towards the South African Constabulary blockhouse line at Langsloot. The Boer doctor was then released, and as he made his way back to the enemy he came across the regimental cows, which had been left by mistake at Springs and were just then catching up the column under the care of a white man who had volunteered to go back for them from Boschman's Kop. A black man also accompanied him. The Boer doctor as he passed remarked to the white man that the cows would shortly be in the possession of the Boers. Sure enough, very soon after, the enemy arrived on the scene and captured the cows. The two men in charge of them were fired at, but managed to escape uninjured.

On Friday, 11 April, the column, with the exception of A and C Squadrons, remained in camp at Langsloot. The two squadrons were, however, moved to Uitkyk and Onverwacht to assist in strengthening the blockhouse line. On the morrow the remainder of the column moved up to the same place for a drive.

14 April.—The column first acted as a stop at Vaal bank on the South African Constabulary blockhouse line in a drive to Val Station southwards, and then moving southwards themselves arrived at that place. The drive resulted in one killed, one wounded, and one hundred and thirty-four captured. Next day after filling up at Val the column trekked to Vlakfontein *via* Greylingstadt, where they halted for lunch. The trek on the morrow took them to Botha's Kraal.

Thursday, 17 April.—They reached the Nigel mine again *via* Heidelberg, and halted to water and feed. They were now in their allotted position for a drive to the north-east which had been arranged to begin at 6 a.m. on the morrow. The drive duly took place and the column reached Balmoral on the 20th. As an operation it was not successful, as only twelve Boers were accounted for. But they had the good fortune to recover their captured cows. They were discovered huddled together on the blockhouse line just where the 7th Hussars finished the drive. Next day they marched from Balmoral to Hollander Pan and camped for the night.

22 April was occupied by a trek to Boschkop; the morrow took them towards Dorstfontein, which they reached on the 24th. They were now in position for another drive. This drive was the seventh down to the Vlakfontein blockhouse line. A trooper of the Royal Scots Greys who had been lost on the previous drive to Balmoral was found in an unconscious state on the *veldt* and died very shortly afterwards. The drive began on Saturday, 26 April, and was continued as far as Strydpan the first night. They finished at Vlakfontein, but the results were nil, such of the enemy as found themselves inside the driven area having managed to escape by way of Springs and Heidelberg. At this time the Royal Scots Greys were with the column.

The 28th was spent at Vlakfontein. Next day the column marched into Heidelberg, their third visit to that place.

Wednesday, 30 April.—The column moved due west to Klip River Station. On the next day they reached Meyerton Station. Friday, 2 May, arrived at Vereeniging and remained there the next day.

On Sunday, 4 May, a draft of one hundred and ten men arrived from England. The column moved at 5.30 a.m. and again crossed the Vaal river into Orange River Colony. They marched to Wolvenhoek Station on the Vereeniging and Kroonstad line and got into position for the eighth drive. The direction was down to the Lindley and Kroonstad blockhouse line. Colonel Rimington was on the right, and Colonel Nixon on the left. The task of the 7th Hussars was to drive over the area bounded by the Heilbron and Vereeniging line on the east, and the Kroonstad and Vereeniging line on the left: these two lines converging at Wolvenhoek.

On 5 May they moved in driving formation to the Rhenoster river, where they camped for the night, forming a strange line from railway line to railway line, wired and entrenched throughout. During the drive A Squadron, 7th Hussars, captured Field Cornet C. B. Prinsloo and twenty-five men.

Next day, orders were received that no wheeled transport or guns were to accompany the column; accordingly all the wagons and guns moved along the Rhenoster river to a station on the Vereeniging-Kroonstad line called Koppes. But A Squadron Cape cart did not. As soon as the wagons had started the column trekked straight through to the Lindley and Kroonstad blockhouse line, a distance of fifty miles. They camped that night at various points along the line. Two squadrons of the 7th Hussars were at Doornkop, while the Bays and A Squadron of the 7th Hussars camped farther west. The total bag for the drive was three hundred and one; the column itself, however, only actually accounted for twenty-five and a few horses.

They remained at Doornkop till 8 May. Towards night on the 8th, in order to shorten the return journey on the morrow, the 7th Hussars marched out a few miles to Bankies. Here they met the remainder of the column from Kroonstad, who, acting on their original orders, had marched into that place on the morning of the 7th. The fresh orders were to remain at Doornkop until the 9th and then to drive back over the old ground to Wolvenhoek in one day.

9 May.—The column started at 6 a.m. and marched fifty miles to Gottenberg on the Heilbron-Vereeniging fine and halted for the night. They returned on the morrow to Grootvlei Station, and were there met by the wagons. The bag for the drive was twenty-eight, of whom the 7th Hussars captured eight. The prisoners were found hiding in the dry bed of a river. They appear to have crossed the Lindley blockhouse line during the first drive and to have recrossed to their old haunts, not expecting a second drive back. It may be remarked that, despite orders to the contrary, the A Squadron Cape cart accompanied the squadron on 5 May and accomplished the entire one hundred and sixty miles in the six days.

The strength of the Regiment at this juncture was five hundred and sixty-two. The casualties up to date were two officers wounded, two non-commissioned officers and men killed, four wounded; three died, and eleven invalided. The events of the remainder of the campaign are absolutely devoid of interest and may be briefly expressed thus—

11 May, Grootvlei; 12 May, Wolvenhoek; 13 May, crossed the Vaal and marched to Engelbrecht Drift; 14 May, moved on to Plat Koppes; and 15 May, Heidelberg. During the Peace Conference, which began on 10 May, there were no hostilities in this district. On 17 May the column returned to Springs and remained there a few days, but owing to a difficulty as to the water supply they returned to Heidelberg, where they remained until the Declaration of Peace on 31 May, 1902.

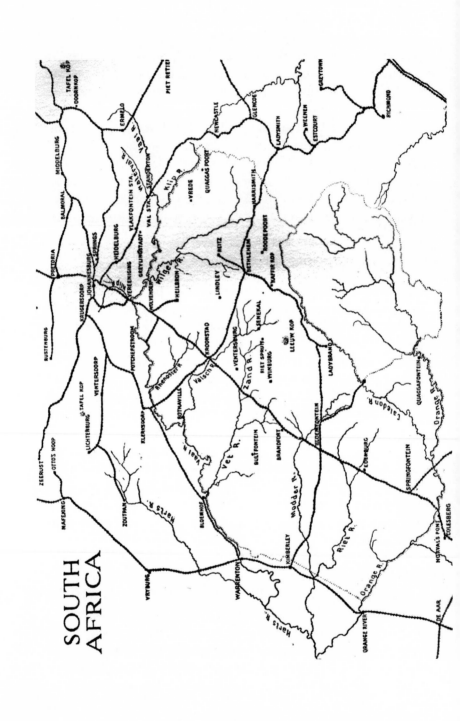

SOUTH
AFRICA

As rewards, Squadron Sergeant-Major Wetherall was awarded the distinguished conduct medal, and Corporal Ketley and Private Tookey were promoted for gallantry.

Peace having been declared, as has been mentioned, the column was broken up. The Regiment marched to Pretoria on 15 June, and camped at Quagga Poort, a site distant six miles from the town.

On 2 July a draft of forty-two men under Lieutenant Kelly joined from England.

The 7th Hussars moved to Krugersdorp on 1 October and camped there. On 25 October a draft of ninety men transferred from the 3rd Hussars joined the Regiment. Another draft of one hundred and fifty-two men, under Lieutenant H.R.H. Prince Arthur of Connaught and Lieutenant the Hon. D. Astley, joined from England on 9 November.

No other event is recorded for the year 1902.

On 19 April Lieutenant A. B. Pollok and Second Lieutenant A. C. Watson, with a draft of two hundred and fifty-nine men, arrived from England. Second Lieutenants E. Brassey and Viscount Maiden joined.

25 June.—Lieut.-Colonel the Hon. R. T. Lawley, C.B., having completed his period of command, was placed on half pay. He was succeeded by Lieut.-Colonel R. L. Walter, who assumed the command on the following day.

The strength of the Regiment on 1 July was as follows: thirty officers, one warrant officer, eight hundred and thirty-six non-commissioned officers and men, and six hundred and twenty-one horses. Though properly belonging to the special volume on Uniform; in which it will be repeated, it may be well here to mention the service dress which was worn by the Regiment during the campaign. The reason for thus inserting it is that it will properly precede the account of the equipment carried on active service and an analysis of the horses contained in a table which shows how they stood the wear and tear of the campaign. The service dress consisted of a khaki serge jacket and trousers, and khaki cord pants and putties. This was taken into use for drill and other duties in January 1902; the undress jacket being abolished, and full dress being reserved for review order, parades, and walking out.

The equipment carried by the Regiment during the war was

as follows: head collar, complete with reins and head rope; Pelham bridle and one rein; saddle, one saddle blanket and one general service blanket; rifle bucket (near side); sword (off side), edge to rear; canvas bandolier on horse's neck; one nosebag on each side of the saddle; off wallet: infantry canteen, groceries in bags, towel and soap; near wallet: Jersey cholera belt, shackle and foot rope, knife, fork and spoon. The rifle was carried in the bucket, bolt outside, safety catch up, and sling round the rider's bridle arm.

The weight of a shoeing smith in marching order was as follows: Rider, 10st. 4½lb.; rifle, 9½lb.; tool-bag (filled), 11lb.; saddle and two blankets, 2st. 31b. feed and   two nose-bags, 9lb.; horse brush and comb, 3lb.; head-dress, 6½lb.; cloak, 9lb.; the total weight, exclusive of ammunition, wallets and their contents, thus being 15st. 13½lb.

From the table which shows how the horses stood the work, and which is based on the number of horses present with the Regiment at Heidelberg on the Declaration of Peace, 31 May 1902, we learn as follows: four hundred and sixty-seven horses were present, out of a total of one thousand and twenty-seven brought from England and received from time to time in South Africa. This four hundred and sixty-seven, however, does not include many good horses left at Kroonstad and Lindley or those in the veterinary hospital at Springs, of which forty-eight would have been fit had it been necessary to march in ten days and ten more in a month's time.—Total, sixty-two.

We will consider the horses squadron by squadron.

A. English, 147 landed, 36 present; De Aar remounts, 72 received, 20 present; Winburg remounts, 3 received, 2 present; Elandsfontein remounts (27.3.02), 30 received, 20 present; remounts (3.5.02), 20 received, 18 present; remounts (16.5.02), 36 received, 36 present; remounts (20.5.02), 11 received, 11 present. B. (omitting the words and merely giving the numbers), 144-34; 69-27; 4-2; 31-23; 27-21; 17-16; 10-10. C. 145-34; 70-20; 8-4; 36-30; 32-30; 26-26; 14-14. Headquarters, 41-10; 20-11; none[1]; 8-6; *2-2;4-4;* none.[1] Totals, 477-114; 231-78; 15-8; 105-79; 81-71; 83-82; 35-35.

At the conclusion of peace twenty-five per cent, of the English horses were still serving and thirty-three per cent, of the remounts received early in January at De Aar.

1. No remounts received at Winburg or on 20 May, 1902.

On 1 May 1903 the Regiment was re-organised on a basis of four Service Squadrons and a Reserve Troop. The new Establishment was as follows: 1 lieut.-colonel; 1 major (second in command); 4 majors; 4 captains; 17 subalterns; 1 adjutant; 1 riding master and 1 quartermaster. Total, 30 officers.

1 regimental sergeant-major and 1 bandmaster. Total, 2 warrant officers. 1 quartermaster sergeant; 1 farrier quartermaster sergeant; 1 sergeant instructor of fencing; 1 squadron sergeant-major, rough rider; 4 squadron sergeants-major; 5 squadron quartermaster sergeants; 1 orderly room sergeant; 1 orderly room clerk, 1 sergeant trumpeter; 1 saddler sergeant; 1 sergeant cook; 1 sergeant master tailor; 5 farrier sergeants and 34 sergeants. Total, 58 sergeants. 9 trumpeters; 34 corporals; 4 shoeing smith corporals, 13 shoeing smiths; 4 saddlers and assistants; 1 saddle-tree-maker and 541 privates. Total, 597 rank and file. Total of all ranks, 696. Horses, 59 officers; 565 riding and 6 draught. Total, 630 horses.

During the remainder of the stay of the Regiment in South Africa no events of importance took place; still, for the purposes of this history we must give a bare list of facts and dates.

10 July.—Inspected at Krugersdorp by Major-General Clements, C.B., D.S.O., Commanding the Pretoria District. On this occasion the medal for 'Distinguished Service in the Field' already mentioned was presented to Squadron Sergeant-Major T. M. E. Wetherall.

Another inspection at the same place was held on 27 and 28 July, the inspecting officer being Brigadier-General J. F. Burn-Murdoch, C.B., Inspector-General of Cavalry, Transvaal.

28 August.—The Regiment proceeded to Klip River Camp for manoeuvres. Here on 3 September General Murdoch again inspected the Regiment. The 7th Hussars returned to Krugersdorp from the manoeuvres on 30 September. Major-General Clements held the annual inspection on 16 October.

One troop of C Squadron under the command of Lieutenant Pollok left Krugersdorp by march route for Potchefstroom on 5 December to be stationed at that place.

9 February, 1904.—The 7th Hussars were inspected at Krugersdorp by Lieut.-General the Hon. Sir N. G. Lyttelton, K.C.B., Commanding the forces in South Africa, on which occa-

Alexander George of Teck

CAPTAIN H. S. H. PRINCE ALEXANDER GEORGE OF TECK, G. C.V. O., D. S. O.

sion he presented the medals gained by the Regiment during the South African War (1901-2).

Inspections were held on 17 June and 4 and 5 July, by Lieut.-General Sir H. J. T. Hildyard. K.C.B., Commanding the forces in South Africa, and Brigadier-General R. G. Broadwood, C.B., Inspector-General of Cavalry, South Africa, respectively.

On 20 July two squadrons, B and D, under the command of Captain H. B. Dalgety, proceeded by march route to Potchefstroom to be stationed there.

The headquarters followed them by train to the same place on 5 August; A and C Squadrons, who proceeded by march route, arriving on 20 August.

Inspections were held on 9 and 10 September and on 14 September by Brigadier-General Broadwood and Brigadier-General Burn-Murdoch respectively.

On 21 September the 7th Hussars proceeded on manoeuvres to a place near Potchefstroom. returning on 27 September.

November 1.—Field-Marshal Earl Roberts, V.C., K.G., who was visiting the station, drove round the lines, a travelling escort being furnished by the 7th Hussars.

By a Special Army Order, promulgated on 21 1 and dated from the War Office, we learn as follows:

His Majesty the King has been graciously pleased to approve of the Regiment being permitted, in recognition of services rendered during the South African War. 1890-1902, to bear upon its colours or appointments the following words: South Africa, 1901-1902.

Inspections were held on 15 February and on 30 and 31 March 1905, by Lieut.-General Hildyard and Brigadier-General Broadwood respectively.

On 1 June 1905 the Establishment of the Regiment was again changed, it being reorganised on a basis of three Service Squadrons and a Reserve Troop. The Establishment now became: 1 lieut.-colonel; 4 majors; 3 captains; 7 lieutenants; 6 second lieutenants; 1 adjutant, 1 riding-master and 1 quartermaster. Total, 24 officers. 1 regimental sergeant-major and 1 bandmaster. Total, 2 warrant officers. 1 quartermaster sergeant, 1 farrier quartermaster sergeant; 1 squadron sergeant-major, rough rider; 1 sergeant instructor in fencing and gymnastics; 1 orderly room sergeant; 3 squadron sergeants-major; 4 squadron quartermaster sergeants, 1 sergeant trumpeter; 1 saddler sergeant; 1 sergeant cook, 1 sergeant master tailor; 4 farrier sergeants, 26 sergeants and 1 orderly room clerk. Total, 47 sergeants. 7 trumpeters; 26 corporals; 3 shoeing-smith corporals; 10 shoeing smiths; 3 saddlers and assistant saddlers; 1 saddle-tree maker and 470 privates. Total, 513 rank and file.

Total of all ranks, 593. Horses, officers, 46; riding, 427; draught, 6. Total, 479.

On 20 and 21 July the Regiment was inspected by Brigadier-General Broadwood. This year the Transvaal manoeuvres were held near Heidelberg, and the 7th Hussars proceeded thither to take part therein on 5 September, returning on 29 September.

Brigadier-General Burn-Murdoch held the annual inspection on 3 October.

On 7 October the station was visited by the Right Hon. the Earl of Selborne, G.C.M.G., High Commissioner of South Africa, on which occasion the Regiment furnished the escort.

The 7th Hussars were now under orders to return home. Accordingly A Squadron and the married families under the command of Major the Hon. J. G. Beresford, D.S.O., proceeded by train to Capetown on 6 November. B and C Squadrons, Headquarters, and the Reserve Troop under command of Lieut.-Colonel R. L. Walter similarly followed the next day.

On 11 November the Regiment, whose strength was twenty-four officers, five hundred and thirty-two non-commissioned officers and men, embarked for England on the hired transport *Dilwara*.

The vessel reached Gibraltar on 29 November, where Captain L. Rawstorne and two sick men were landed. At Gibraltar Captain Rawstorne, who was suffering from enteric, died on 4 December.

The *Dilwara* reached Southampton on 5 December.

CHAPTER 12

Towards the Great War

On arrival at Southampton the 7th Hussars, with the exception of A Squadron, proceeded by train to Norwich, but A Squadron entrained for Weedon. Being now in England the Regiment was placed on the Home Establishment—three Service Squadrons and a Reserve Squadron. This involved the following changes: four captains instead of three; eight lieutenants instead of seven; the riding-master was temporary; a squadron sergeant-major instructor in musketry was added, as was another squadron sergeant-major, and three more sergeants. This gave a total of twenty-four officers, two warrant officers, and fifty-two sergeants; an eighth trumpeter was added, and four corporals, one shoeing-smith and one saddler and assistant saddler; and there were five hundred and seventy-eight privates instead of four hundred and seventy. Thus the total rank and file became six hundred and twenty-seven, and the total of all ranks seven hundred and fifteen. Officers' horses fifty, riding horses four hundred and fifty-nine, and draught horses six.

On 9 December 1905, Major-General F.W. Benson. C.B., Director of Transport and Remounts, inspected the horses of the Regiment.

15 December.—A memorial service was held in Norwich Cathedral to the late Captain L. Rawstorne and Private J. Reid, who died at Gibraltar.

The events of the year 1906 were as follows—

27 January.—Barracks inspected by General Lord Methuen, G.C.B., K.C.V.O., C.M.G., Commander-in-Chief Eastern Command.

28 January.—Report received from Major-General T. E. Stephenson, C.B., Commanding the Transvaal District.

7th Hussars: This Regiment is in a thoroughly satisfactory condition and is in every way fit for active service.

14 February.—One hundred recruits proceeded to Colchester for musketry.

27 February.—The barracks and a draft for India were inspected by Brigadier-General Allenby, C.B., Commanding the 4th Cavalry Brigade. This draft, which was posted to the 3rd Hussars, embarked for India per hired transport *Ionian* on 28 *February*.

The barracks were again inspected on 8 March, this time by Major-General Sir Edward Hutton, K.C.M.G., C.B.

13 March.—Another party of seventy recruits proceeded to Colchester for musketry.

2 April.—A draft of twenty-nine non-commissioned officers and men arrived from York and were posted to the Regiment from the 3rd Hussars.

The annual inspection of the Regiment by Brigadier-General Allenby was held on 25 April.

22 May.—The officers of the Regiment attended his Majesty's levee on return from foreign service.

26 May.—Nine officers and three hundred and eighteen non-commissioned officers and men, under the command of Major C. E. G. Norton, proceeded to Landguard by rail for the annual course of musketry, returning to Norwich on 22 June. Next day seven officers and two hundred and twenty-nine non-commissioned officers and men under the command of Major R. M. Poore, D.S.O., similarly went to Landguard, and returned on 19 July.

11 July.—The Regiment was completed to Establishment with horses from the 14th Hussars.

27 July.—Major-General R. S. S. Baden-Powell, C.B., Inspector of Cavalry, held an inspection of the Regiment.

5 September.—A draft of forty-five men embarked for India per hired transport *Assaye* for posting to the 3rd Hussars.

29 October.—General Lord Methuen, G.C.B., K.C.V.O., C.M.G. Commander-in-Chief Eastern Command, visited Norwich, saw the recruits at riding school and inspected the barracks.

On 18 December 1906 and again on 22 January and on 1 and 26 February 1907, Brigadier-General Allenby, C.B., inspected the regiment at winter training.

LIEUTENANT H. R. H. PRINCE ARTHUR OF CONNAUGHT, K. G., G. C.V. O.

16 February.—Seventy-one recruits proceeded to Colchester for musketry.

9 March.—A draft of twenty-five men embarked for India per hired transport *Assaye,* for posting to the 3rd Hussars.

27 March.—The horses of the Regiment were inspected by Major-General Benson, C.B.

2 April.—The War Office having decided to withdraw all quick-firing equipment (pom-pom) for cavalry regiments, the gun and equipment were returned to Ordnance.

27 April.—Lieutenant His Royal Highness Prince Arthur of Connaught, K.G., K.C.V.O., personal *aide-de-camp* to the King, was promoted Captain in the 2nd Dragoons (Royal Scots Greys).

4 May.—The Regiment was inspected by Major-General R. S. S. Baden-Powell, C.B.. Inspector of Cavalry.

12 May.—This day the Colonel of the Regiment, Major-General Robert Hale, died.

He was succeeded in the colonelcy of the 7th Hussars by Major-General Sir Hugh McCalmont, K.C.B., C.V.O., 13 May 1907.

27 May.—General Lord Methuen inspected the Regiment on Mousehold Heath.

The annual inspection was held by Brigadier-General Allenby on 10 June.

During May and June the Regiment proceeded by rail to Landguard for the annual course of musketry in two parties; the first half going upon 27 May, and the remainder on 24 June.

26 June.—Colonel R. L. Walter, having completed his period of command of the Regiment, was placed on half pay. He was succeeded by Colonel G. L. Holdsworth.

10 July.—An inspection was held by Major-General H. J. Scobell, C.B., Inspector of Cavalry.

25 July.—It being the season for brigade training and manoeuvres, the Regiment proceeded by march route to Bulford Camp, where they arrived on 3 August.

En route the 7th Hussars camped at Diss, Ipswich, Colchester, Chelmsford, Woolwich, Hounslow, Aldershot, and Overton.

The Regiment proceeded from Bulford Camp to Aldershot by road on 3 October, to be stationed there; the details from Norwich arriving by train on the same day.

228

8 November.—The band and a complete squadron proceeded by road to London *via* Hounslow for duty at the Lord Mayor's Show.

The party remained to be employed on street duty on the occasion of the visit of his Imperial Majesty the German Emperor when he visited the city on 13 November.

31 January 1908.—Inspection of the troops at winter training by Major-General Scobell.

1 February.—A draft of eighty-one non-commissioned officers and men proceeded to Southampton for embarkation on s.s. *Assaye* for posting to the 3rd Hussars in South Africa.

2 April.—The horses of the Regiment were inspected by Major-General Scobell.

18 May.—Field operations on the Long Valley in presence of his Majesty the King.

20 May.—The regiment, as part of the 1st Cavalry Brigade, was inspected on the Long Valley by H.R.H. the Prince of Wales, now King George V.

On the same day their Royal Highnesses the Prince and Princess of Wales visited the officers' mess and afterwards witnessed a display in the riding school.

11 June.—Inspection of the horses by Colonel D. E. Wood, Inspector of Remounts.

26 June.—Review on Laffan's Plain on the occasion of the celebration of his Majesty the King's birthday.

10 July.—Annual inspection by Major-General the Hon. J. N. G. Byng, C.B., M.V.O., Commanding the 1st Cavalry Brigade.

13 July.—Inspection by Major-General H. J. Scobell, C.V.O., C.B.

11 August.—Cavalry Divisional manoeuvres on Salisbury Plain. The regiment returned to Aldershot on 30 August.

4 September.—Horses inspected by Lieut.-General Sir H. L. Smith-Dorrien, K.C.B., D.S.O., Commanding-in-Chief Aldershot Command.

14 September.—Manoeuvres of the Aldershot Command in the neighbourhood of Winchester. The Regiment returned on 19 September.

22 September.—Institutes, messes, &c, inspected by Lieut.-General Sir H. S. Smith-Dorrien.

ALDERSHOT, 1910

31 October.—Eighty-eight non-commissioned officers and men proceeded to Southampton for embarkation to South Africa, for posting to the 3rd Hussars.

24 February 1909.—A similar draft (eighty-six men) was also sent to the 3rd Hussars in South Africa.

13 May.—Field operations in the neighbourhood of Elstead in the presence of his Royal Highness the Prince of Wales.

18 May.—Field operations on the Chobham Ridges in the presence of his Majesty the King.

11 June.—On the occasion of the visit of the Imperial Press Delegates and Foreign Officers the Regiment took part in field operations near Pirbright.

1 July.—In an army order dated 1 July 1909, his Majesty the King was graciously pleased to approve of the regiment being permitted, in recognition of services rendered at the following battle, to bear upon its appointments the following distinction: Warburg (Battle of Warburg, 31 July 1760)

1 August.—In an Army Order dated 1 August 1909, his Majesty the King was graciously pleased to approve of the regiment being permitted, in recognition of services rendered at the following battles, to bear on its appointments the undermentioned distinctions: Beaumont (Battle of Beaumont, 26 April 1794); Willems (Battle of Willems, 10 May 1794).

26 August.—Cavalry divisional army and command manoeuvres. The Regiment proceeded to Salisbury Plain by road and returned to Aldershot on 24 September.

20 October.—A draft of one hundred and five men under command of Second Lieutenant Sir T. E. K. Lees, Bart., 15th Hussars, embarked for South Africa per steamship *Braemar Castle* for posting to the 3rd Hussars.

3 November.—Barracks inspected by Lieut.-General Sir H. E. Smith-Dorrien, K.C.B., D.S.O., Commanding-in-Chief Aldershot Command.

15 November.—In a War Office letter No. 61913, dated 15 November 1909, his Majesty the King was graciously pleased to approve the letters Q O interlaced within the Garter, being borne upon the appointments of the Regiment in place of the Royal Cypher within the Garter. The blue and gold original official badge is

preserved in the War Office library. This differs, as will be seen, from the unofficial though time-honoured badge of Q. O. interlaced and surmounted by a queen consort's crown. Of this old badge several variants are given in other places in the book.

6 May 1910.—The lamented death of his Majesty King Edward the Seventh took place at Buckingham Palace.

20 May.—On the occasion of the funeral of his late Majesty, a detachment of one hundred non-commissioned officers and men of the regiment under the command of Major J. Fryer proceeded by rail to London for duty (dismounted) in lining the streets. The Regiment also furnished twenty horses (bays) to mount the detachment of the 10th Hussars taking part in the procession.

11 July.—Their Majesties King George the 5th and Queen Mary arrived at Aldershot and stayed at the Royal Pavilion until 16 July.

14 July.—The 1st Cavalry Brigade, consisting of the 3rd Dragoon Guards, 7th Hussars, and 19th Hussars, were inspected by his Majesty the King on Long Valley.

5 August.—Annual inspection by Brigadier-General C. T. Kavanagh, C.B., C.V.O., D.S.O.

25 August.—The regiment proceeded by road to Salisbury Plain for the Cavalry Division command and army manoeuvres.

21 September.—The details left behind at Aldershot (strength, four officers, two hundred and thirty-eight non-commissioned of-

WARRANT OFFICERS, STAFF SERGEANTS & SERGEANTS 7TH (Q. O.) HUSSARS, HOUNSLOW, JUNE 1911

ficers and men, and one hundred and sixty-seven horses), under the command of Major J. Fryer, proceeded by road and rail to Hounslow on change of station.

27 September.—The Regiment (strength, nineteen officers, three hundred and fifty-five non-commissioned officers and men, and three hundred and two horses), under the command of Colonel G. L. Holdsworth, arrived at Hounslow from the manoeuvres.

29 October.—Inspection of horses by Brigadier-General F. S. Garratt, C.B., D.S.O., Commanding 4th Cavalry Brigade.

9 November.—The Regiment furnished a detachment for duty in London at the Lord Mayor's Show.

8 December.—Inspection at winter training by Brigadier-General Garratt, C.B., D.S.O.

21 December.—A draft of twenty-five men under under the command of Lieutenant R. R. Grubb, 3rd Hussars, proceeded to Southampton for embarkation to South Africa per s.s. *Soudan* for posting to the 3rd Hussars.

12 January 1911.—Inspection by Major-General Allenby, C.B Inspector of Cavalry.

26 January.—Inspection by Lieut.-General Sir A. H. Paget K.C.B., K.C.V.O., A.D.C., Commanding-in-Chief the Eastern Command.

6 February.—The Regiment proceeded to London for duty in lining the streets on the occasion of the opening of Parliament by his Majesty the King.

15 March.—Inspection by Brigadier-General F. S. Garratt, C.B. D.S.O.

26 June, 1911.—Lieut.-Colonel R. M. Poore, D.S.O., succeeded to the command of the Regiment.

The fifth annual dinner of the 7th (Queen's Own) Hussars Old Comrades' Association was held at the King's Hall, Holborn Restaurant, on the evening of Saturday, 11 October 1913.

Upwards of one hundred and sixty past and present members of the Regiment spent a most enjoyable evening together on the occasion. The Colonel of the Regiment, Major-General Sir Hugh McCalmont, K.C.B., C.V.O., presided, supported by Major-General Sir H. A. Bushman, K.C.B., Colonel Harold Paget, C.B., D.S.O., Colonels Hunt, Ridley, Nicholson, Holdsworth and Walter; Majors Fryer, Norton and Hermon; Captains McCalmont, Bates, Leyland, and Brassey. Greetings were received from the Regiment at Bangalore. The loyal toasts were duly honoured, and a special toast for 'Prince Arthur of Connaught and the Duchess of Fife,' whose marriage was so soon to take place, was given; a message of congratulation being also forwarded to H.R.H. Prince Arthur, who was formerly an officer of the Regiment. Sir Hugh McCalmont proposed 'The Regiment,' to which Major Norton, second in command, replied, and in his speech incidentally mentioned the success of Sergeant-Major Webb at Olympia this year, where he won a gold medal, and also the victories of the Regiment at the South of India Rifle Meeting. Other toasts followed, among which that of 'The Health of the Colonel, Sir Hugh McCalmont,' was enthusiastically received. The Old Comrades' Association of the Regiment is doing a good work, and will, it is to be hoped, continue to prosper.

An interesting point about the meeting may be noted, the presence of the Old Comrades' Association of the Lincolnshire Regiment in the building. where their 10th annual dinner was being held in the Throne Room. This becoming known, a deputation from the 7th (Queen's Own) Hussars proceeded thither to carry greetings from the regiment to the Lincolnshires.

It is gratifying to learn that the Old Comrades' Association of the 7th Hussars is in a nourishing condition. May it be even more so when next year comes round and the Old Comrades meet again!

In *October* 1911 the Regiment was ordered to sail for India, its destination being Bangalore.

It will be observed that the stay at home of the 7th (Queen's Own) Hussars was more than usually short. The Regiment is still stationed at Bangalore, February 1914.

LEONAUR

ALSO FROM LEONAUR
AVAILABLE IN SOFTCOVER OR HARDCOVER WITH DUST JACKET

WAR BEYOND THE DRAGON PAGODA by *J. J. Snodgrass*—A Personal Narrative of the First Anglo-Burmese War 1824 - 1826.

ALL FOR A SHILLING A DAY by *Donald F. Featherstone*—The story of H.M. 16th, the Queen's Lancers During the first Sikh War 1845-1846.

AT THEM WITH THE BAYONET by *Donald F. Featherstone*—The first Anglo-Sikh War 1845-1846.

A LEONAUR ORIGINAL

THE HERO OF ALIWAL by *James Humphries*—The days when young Harry Smith wore the green jacket of the 95th-Wellington's famous riflemen-campaigning in Spain against Napoleon's French with his beautiful young bride Juana have long gone. Now, Sir Harry Smith is in his fifties approaching the end of a long career. His position in the Cape colony ends with an appointment as Deputy Adjutant-General to the army in India. There he joins the staff of Sir Hugh Gough to experience an Indian battlefield in the Gwalior War of 1843 as the power of the Marathas is finally crushed. Smith has little time for his superior's 'bull at a gate' style of battlefield tactics, but independent command is denied him. Little does he realise that the greatest opportunity of his military life is close at hand.

THE GURKHA WAR by *H. T. Prinsep*—The Anglo-Nepalese Conflict in North East India 1814-1816.

SOUND ADVANCE! by *Joseph Anderson*—Experiences of an officer of HM 50th regiment in Australia, Burma & the Gwalior war.

THE CAMPAIGN OF THE INDUS by *Thomas Holdsworth*—Experiences of a British Officer of the 2nd (Queen's Royal) Regiment in the Campaign to Place Shah Shuja on the Throne of Afghanistan 1838 - 1840.

WITH THE MADRAS EUROPEAN REGIMENT IN BURMA by *John Butler*—The Experiences of an Officer of the Honourable East India Company's Army During the First Anglo-Burmese War 1824 - 1826.

BESIEGED IN LUCKNOW by *Martin Richard Gubbins*—The Experiences of the Defender of 'Gubbins Post' before & during the sige of the residency at Lucknow, Indian Mutiny, 1857.

THE STORY OF THE GUIDES by *G.J. Younghusband*—The Exploits of the famous Indian Army Regiment from the northwest frontier 1847 - 1900.

Lightning Source UK Ltd.
Milton Keynes UK
UKOW02f2019031215

264072UK00001B/183/P